# The essence of chromatography

# The essence of chromatography

Colin F. Poole

*Department of Chemistry, Wayne State University, Detroit, MI 48202, USA*

**2003**
**ELSEVIER**
**Amsterdam – Boston – London – New York – Oxford – Paris**
**San Diego – San Francisco – Singapore – Sidney – Tokyo**

ELSEVIER SCIENCE B.V.
Radarweg 29, PO Box 211, 1000 AE Amsterdam, The Netherlands



First edition 2003

Library of Congress Cataloging in Publication Data
A catalog record from the Library of Congress has been applied for.

British Library Cataloguing in Publication Data
A catalogue record from the British Library has been applied for.

ISBN: 0 444 50198 3 (hardbound)
ISBN: 0 444 50199 1 (paperback)

Transferred to digital print 2008
Printed and bound in Great Britain by CPI Antony Rowe, Eastbourne

# Contents

# Preface

The knowledge base of chromatography continued to expand throughout the 1990s owing to its many applications to problems of contemporary interest in industry and life and environmental sciences. Organizing this information into a single text for a diverse group of scientists has become increasingly difficult. The present book stemmed from the desire to revise an earlier work, "Chromatography Today", published in 1991. It was soon realized that a simple revision would not provide the desired result of a contemporary picture of the practice of chromatography at the turn of the century. The only workable solution was to start afresh, maintaining the same general philosophy and concept for "Chromatography Today" where possible, while creating essentially a new book. In particular, both time and space constraints dictated that to cover in equal depth the diverse separation techniques in current use, it would not be possible to cover sample preparation techniques to the same extent as "Chromatography Today". The division I made here was to include automated and on-line methods with instrumentation, and treat them in a comprehensive manner, while widely used manual laboratory operations are not treated at all, albeit that these techniques are an integral part of laboratory life. This allowed, for example, the addition of a comprehensive and separate chapter on capillary-electromigration separation techniques, and greater emphasis on modern approaches for data analysis, compared with "Chromatography Today".

In writing this book, I had in mind that it would present a comprehensive survey of modern chromatographic and capillary electrophoretic techniques at a level commensurate with the needs of a textbook for teaching post-baccalaureate courses in the separation sciences. In addition, it would fulfill the need for a self-study guide for professional scientists wishing to refresh their background in this rapidly growing field. The chapters follow a modular format to allow instructors to select components to their liking to make up a typical one-semester course. For the professional scientist, the extensive cross-referencing and comprehensive index should allow individual topics to be quickly found, and the extensive bibliography to be used for entry into the primary scientific literature. Where possible, frequently searched for characteristic properties of separation systems are collected in tables, to allow the book to be used as a stand-alone resource for the professional scientist.

Colin F. Poole

# Chapter 1

# General Concepts in Column Chromatography

## 1.1 INTRODUCTION

The Russian botanist M. S. Tswett is generally credited with the discovery of chromatography around the turn of the century [1,2]. He used a column of powdered calcium carbonate to separate green leaf pigments into a series of colored bands by allowing a solvent to percolate through the column bed. He also coined the name chromatography (color writing) from the Greek for color (chroma) and write (graphein) to describe the process. However, column liquid chromatography as described by Tswett was not an instant success, and it was not until its rediscovery in the early 1930s that it became an established laboratory method. Chemists at this time were limited to such laboratory tools as crystallization, liquid-liquid distribution and distillation for separations and new techniques were needed for the rapid isolation of pure components from natural products and to support the development of increasingly sophisticated approaches to organic synthesis. Although many scientists made substantial contributions to the evolution of modern chromatography, not least among these is A. J. P. Martin who received the Nobel prize in 1952 for the invention of partition chromatography (with R. L. M. Synge) and in the same year with A. T. James he introduced the technique of gas-liquid chromatography. The 1940s saw a rapid expansion in the use of chromatographic methods in the laboratory but the introduction and development of gas-liquid chromatography in the 1950s represented a significant milestone, ushering in the era of instrumental methods of separation which spawned the many variations of modern chromatography in use today. Further milestones in the evolution of chromatographic separation methods are summarized in Table 1 [3-5]. Individual profiles of the early pioneers of chromatography are collected in ref. [6-8].

## 1.2 FAMILY TREE OF CHROMATOGRAPHIC METHODS

Since chromatography has evolved into a large number of applied methods it is no simple task to provide a meaningful comprehensive definition. Chromatography is essentially a physical method of separation in which the components to be separated are distributed between two phases, one of which is stationary (stationary phase) while the other (the mobile phase) moves in a definite direction [9,10]. This definition suggests that chromatographic separations have three distinct features: (a) they are physical methods of separation; (b) two distinct phases are involved, one of which is stationary while the other is mobile; and (c) separation results from differences in the distribution constants of the individual sample components between the two phases. The definition could be broadened to allow for the fact that it is not essential that one phase is stationary, although this may be more experimentally convenient. What is important, is that either the rate of migration or direction of migration of the two phases are different [11]. Micellar electrokinetic chromatography (MEKC) is an example of a separation technique based on differential migration in a two-phase system. The above definition excludes all separations that occur by differential migration in a

Table 1.1
Some significant time frames in the evolution of modern chromatography

| Date | Associated development |
|---|---|
| 1903 | • Original description of column liquid chromatography by Tswett |
| 1931 | • Column liquid chromatography rediscovered by Lederer and co-workers at a time more receptive for its establishment as a standard laboratory method. |
| 1938 | • Ion-exchange column liquid chromatography introduced. It came to prominence as a distinct chromatographic technique during the Second World War (1939-1945) as a separation procedure for the rare earth and transuranium elements. |
| 1941 | • Column liquid-liquid partition chromatography introduced as a faster and more efficient separation method than countercurrent distribution chromatography. |
| 1944 | • Paper chromatography introduced as a fast, simple and convenient method for analytical separations based on partition chromatography. Now largely replaced by thin-layer chromatography. |
| Mid-1940s | • Gel electrophoresis developed for the separation of charged analytes in a stabilizing gel matrix. Later became an important method for the separation of biopolymers. |
| Early-1950s | • Immobilized layers and standardized sorbents leads to the popularization of thin-layer chromatography as a faster and more convenient method than column liquid chromatography for analytical separations. Fine-particle layers introduced in the mid-1970s were responsible for the development of high performance (instrumental) thin-layer chromatography. |
| 1952 | • Gas-liquid chromatography is described by James and Martin and begins the development of instrumental chromatographic methods. Gas chromatography provided a major improvement in the separation of volatile compounds eclipsing established methods at that time. It remains the most widely used chromatographic technique for the fast and efficient separation of thermally stable and volatile compounds today. |
| 1958 | • Column liquid size-exclusion chromatography using controlled porosity dextran gels introduced by Flodin and Porath. This became an important approach for the separation (or characterization) of polymers based on size differences as well as for the estimation of molecular weights. |
| 1962 | • Klesper introduced supercritical fluids as mobile phase for column chromatography but limited development took place until the early 1980s when Lee introduced open tubular columns. Most supercritical fluid separations today use packed columns of small internal diameter. |
| Mid-1960s | • Giddings introduces the technique of field flow fractionation for the separation of particles and continues to develop the theory and technology of its many variants (fields) over the next 30 years. |
| 1967 | • Affinity chromatography introduced by Porath and co-workers for the isolation of biological polymers based on the specificity of their interactions with appropriate immobilized ligands. |
| Late-1960s | • The introduction of pellicular sorbents catalyzed the development of high pressure liquid chromatography. It was not until the mid-1970s that rapid development took place with the introduction of porous microparticle sorbents. By the 1980s high pressure liquid chromatography was well established as the most popular condensed phase separation technique in modern chromatography. |
| 1970 | • Everaerts and co-workers introduced capillary isotachophoresis for the concentration and separation of ions. |
| 1970s | • Ito and co-workers commenced a number of advances in counter current chromatography using centrifugal and planetary motion for liquid-liquid separations. |
| Mid-1970s | • Small and co-workers introduced ion chromatography based on the integration of ion-exchange chromatography with conductivity detection for the analysis of ions. This method is now the most common chromatographic technique for the analysis of inorganic ions. |

Table 1.1
(Continued)

| Date | Associated development |
|---|---|
| Early-1980s | • Jorgenson and co-workers popularized the use of zone electrophoresis in capillary columns for the fast and efficient separation of ions and biopolymers. |
| 1984 | • Terabe introduced the method of micellar electrokinetic chromatography (MEKC) using surfactant-containing buffers in a capillary electrophoresis apparatus. Over the next decade MEKC matured into an important method for the electroseparation of neutral compounds. |
| Late-1980s | • Rediscovery of capillary electrochromatography. Pioneering work by Knox leads to the evolutionary development of this technique during the 1990s. |

single-phase system, such as capillary electrophoresis (CE). Useful chromatographic separations require an adequate difference in the strength of physical interactions for the sample components in the two phases, combined with a favorable contribution from system transport properties that control the movement within and between phases. Several key factors are responsible, therefore, or act together, to produce an acceptable separation. Individual compounds are distinguished by their ability to participate in common intermolecular interactions in the two phases, which can generally be characterized by an equilibrium constant, and is thus a property predicted from chemical thermodynamics. During transport through or over the stationary phase differential transport resulting from diffusion, convection, turbulence, etc., result in dispersion of solute zones around an average value, such that they occupy a finite distance along the stationary phase in the direction of migration. The extent of dispersion restricts the capacity of the chromatographic system to separate, and independently of favorable thermodynamic contributions to the separation, there are a finite number of dispersed zones that can be accommodated in the separation. Consequently, chromatographic separations depend on a favorable contribution from thermodynamic and kinetic properties of the compounds to be separated.

A convenient classification of chromatographic techniques can be made in terms of the physical state of the phases employed for the separation, Figure 1.1. When the mobile phase is a gas and the stationary phase a solid or liquid the separation techniques are known as gas-solid chromatography (GSC) or gas-liquid chromatography (GLC), respectively. Gas-liquid chromatography is the more popular separation mode and is often simply referred to as gas chromatography (GC). When the mobile phase is a supercritical fluid and the stationary phase either a solid or immobilized liquid the separation technique is called supercritical fluid chromatography (SFC). For gas and supercritical fluid chromatography the dominant separation mechanisms are partitioning between bulk phases and interfacial adsorption. To classify separation techniques with a liquid mobile phase a wider range of separation mechanisms needs to be considered and is commonly used as the basis of classification. Also, true liquid-liquid separation systems are not important because of their limited stability and experimental inconvenience. Modern liquid chromatography is dominated by the use of inorganic oxides with organic functional groups chemically bonded to their surface, known as bonded phases, and to a lesser extent porous polymers. When the stationary phase is

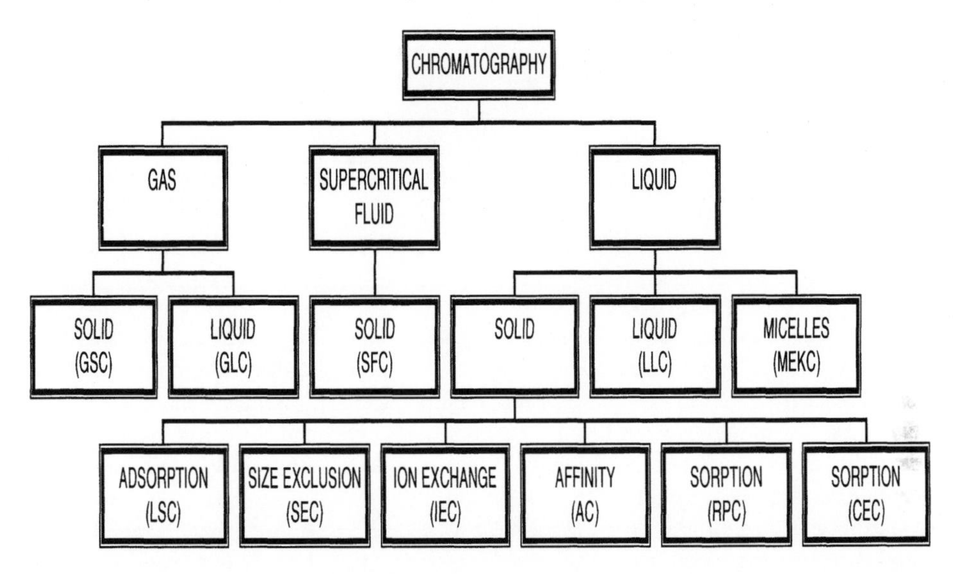


Figure 1.1. Family tree of column chromatographic methods. GSC = gas-solid chromatography; GLC = gas-liquid chromatography; SFC = supercritical fluid chromatography; LLC = liquid-liquid chromatography; MEKC = micellar electrokinetic chromatography; LSC = liquid-solid chromatography; SEC = size- exclusion chromatography; IEC = ion-exchange chromatography; AC = affinity chromatography; RPC = reversed-phase chromatography; and CEC = capillary electrochromatography.

a solid and interfacial adsorption the dominant separation mechanism the technique is referred to as liquid-solid chromatography (LSC). If the stationary phase is a solid with a controlled pore size distribution and solutes are separated by size differences then the technique is referred to as size-exclusion chromatography (SEC). If the stationary phase is a solid with immobilized ionic groups and the dominant separation mechanism is electrostatic interactions between ions in the mobile phase and those on the stationary phase then the technique is referred to as ion-exchange chromatography (IEC) or ion chromatography (IC). If the stationary phase is a solid with immobilized molecular recognition sites in which the dominant separation mechanism is the three-dimensional specificity of the interaction between the molecular recognition site and the sample then the technique is referred to as affinity chromatography (AC). Reversed-phase chromatography (RPC) is a particular form of bonded-phase chromatography in which the mobile phase is more polar than the stationary phase (for most practical applications the mobile phase is an aqueous solution). Reversed-phase chromatography is the most popular separation mode in modern liquid chromatography being applicable to a wide range of neutral compounds of different polarity. In addition, by exploiting secondary chemical equilibria in the mobile phase, ionic compounds are easily handled by ion suppression, ion pairing, or complexation.

In the normal operating mode in gas, supercritical fluid and liquid chromatography the stationary phase is contained in a rigid container, usually a tube of various dimensions, called a column, through which the mobile phase is forced to migrate by external pressure. Alternatively, the bulk flow of mobile phase containing an electrolyte can be induced by an external electric field through the process known as electroosmosis. When a column containing a stationary phase is used and the movement of the mobile phase is caused by electroosmosis the separation technique is called electrochromatography, or since columns of capillary dimensions are essential for this technique, capillary electrochromatography (CEC). Ionic surfactants can form micelles as a continuous phase dispersed throughout a buffer. In an electric field these charged micelles move with a different velocity or direction to the flow of bulk electrolyte. Neutral solutes can be separated if their distribution constants between the micelles and buffer are different by the technique known as micellar electrokinetic chromatography (MEKC). Ionic solutes in CEC and MEKC are influenced by the presence of the electric field and are separated by a combination of chromatography and electrophoresis. All the above processes are considered examples of column chromatography. If the stationary phase is distributed as a thin layer on a (usually) flat support, such as a sheet of glass or plastic, and the mobile phase is allowed to ascend through the layer (usually) by capillary forces then this method is referred to as planar or thin-layer chromatography (TLC). TLC has largely replaced paper chromatography (PC) in contemporary practice owing to the poorer separation characteristics of the latter.

## 1.3 ZONE MIGRATION

Transport of solute zones in column chromatography occurs entirely in the mobile phase. Transport is an essential component of the chromatographic system since the common arrangement for the experiment employs a sample inlet and detector at opposite ends of the column with sample introduction and detection occurring in the mobile phase. There are three basic approaches for achieving selective zone migration in column chromatography, Figure 1.2 [12]. In frontal analysis, the sample is introduced continuously onto the column as a component of the mobile phase. Each solute is retained to a different extent as it reaches equilibrium with the stationary phase until, eventually, the least retained solute exits the column followed by other zones in turn, each of which contains several components identical to the solutes in the zone eluting before it [13]. Ideally the detector output will be comprised of a series of rectangular steps of increasing height. Frontal analysis is used to determine sorption isotherms for single component or simple mixtures and to isolate a less strongly retained trace component from a major component. Quantification of each component in a mixture is difficult and at the end of the experiment, the column is contaminated by the sample. For these reasons frontal analysis is used only occasionally for separations. Frontal analysis is the basis of solid-phase extraction techniques used for the collection of contaminants from air and water by sorption onto short sorbent beds.

In displacement chromatography the sample is applied to the column as a discrete band and a substance (or mobile phase component) with a higher affinity for

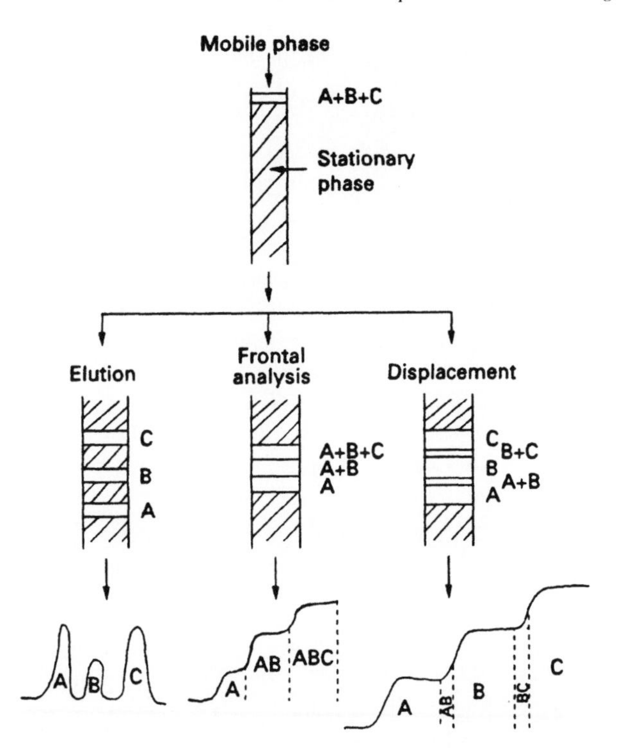


Figure 1.2. Mode of zone displacement in column chromatography.

the stationary phase than any of the sample components, called the displacer, is continuously passed through the column. The displacer pushes sample components down the column, and if the column is long enough, a steady state is reached, and a succession of rectangular bands of pure components exit the column. Each component displaces the component ahead of it, with the last and most strongly retained component being forced along by the displacer. At the end of the separation the displacer must be stripped from the column if the column is to be reused. Displacement chromatography is used mainly in preparative and process chromatography, where high throughputs of pure compounds can be obtained (section 11.3.5) [12]. Depending on the experimental conditions the contact boundary between zones may not be discrete and the collection of pure material may be restricted to the central region of the displaced zones.

In elution chromatography the mobile and stationary phase are normally at equilibrium. The sample is applied to the column as a discrete band and sample components are successively eluted from the column diluted by mobile phase. The mobile phase must compete with the stationary phase for the sample components and for a separation to occur the distribution constants for the sample components resulting from the competition must be different. Elution chromatography is the most convenient method for analysis and is commonly used in preparative-scale chromatography.

Today elution development has become synonymous with the word chromatography itself.

The information obtained from a chromatographic experiment is contained in the chromatogram. When the elution mode is used this consists of a plot of (usually) detector response (y-axis) as a continuous function of time or volume of mobile phase passed through the column (x-axis). The chromatogram contains a number of peaks of various sizes rising from a baseline. Many representative examples can be found throughout this text. Information readily extracted from the chromatogram includes an indication of sample complexity from the number of observed peaks; qualitative identification of sample components from the accurate determination of peak position; quantitative assessment of the relative concentration or amount of each component from their peak areas; and characteristic physical properties of either the solute or the chromatographic system from peak positions and profiles. The fundamental information of the chromatographic process that can be extracted from the chromatogram forms basis of the remainder of this chapter.

## 1.4 RETENTION

The position of a peak in a chromatogram is characterized by its retention time ($t_R$) or retention volume ($V_R$). Retention volumes are fundamentally more correct than time but require further experimental information for their determination. We will come to this shortly, and consider only the directly observable measurement of time for the present. The retention time is made up of two components. The time that the solute spends in the mobile phase and the time it spends in the stationary phase. All solutes spend the same time in the mobile phase, which is simply the time required by an unretained solute, that is a solute that does not interact with the stationary phase, to travel through the chromatographic system. This time is called the column hold-up time, $t_M$, (sometimes referred to as the dead time although hold-up time is preferred). It represents the time required by the mobile phase entering the column to reach the detector and in volume terms is equivalent to the volume of streaming mobile phase contained in the column. In liquid and supercritical fluid chromatography a fraction of the mobile phase can be trapped in the pores of the column packing and is stagnant. The volume of stagnant mobile phase is considered part of the stationary phase and thus the column hold-up volume is less than the volume of liquid or fluid filling the column. For a gas the column hold-up volume and the unoccupied volume of the column are identical. The time the solute spends in the stationary phase is called the adjusted retention time, $t_R'$ (or adjusted retention volume, $V_R'$) and is calculated by difference from the retention time (volume) and the column hold-up time (volume). Since for convenience the retention time of a substance is determined from the moment of injection as time zero, we arrive at the simple relationship (Eq. 1.1) combining the independent contributions to the observed retention time

$$t_R = t_M + t_R' \qquad (1.1)$$

For the optimization of chromatographic separations and in the formulation of theoretical models the retention factor (sometimes referred to as the capacity factor), k, is more important than retention time. The retention factor is the ratio of the time a substance spends in the stationary phase to the time it spends in the mobile phase (Eq. 1.2)

$$k = t_R' / t_M = (t_R - t_M) / t_M \quad (1.2)$$

If the distribution constant is independent of the sample amount then the retention factor is also equal to the ratio of the amounts of substance in the stationary and mobile phases. At equilibrium the instantaneous fraction of a substance contained in the mobile phase is $1 / (1 + k)$ and in the stationary phase $k / (1 + k)$. The retention time and the retention factor are also related through Eq. (1.3)

$$t_R = t_M (1 + k) = (L / u) (1 + k) \quad (1.3)$$

where L is the column length, and u the average mobile phase velocity. The distribution constant for a substance in the chromatographic system is equal to the product of the retention factor and the phase ratio ($K = k\beta$). The phase ratio, $\beta$, is defined as the ratio of the volume of mobile phase and stationary phases in the column for a partition system, or the ratio of the volume of the mobile phase to the surface area of the stationary phase for an adsorption system, respectively.

The relative retention of any two peaks in the chromatogram is described by the separation factor, $\alpha$, given by Eq. (1.4)

$$\alpha = t_R'_{(B)} / t_R'_{(A)} = k_B / k_A \quad (1.4)$$

By convention, the adjusted retention time or retention factor of the later eluting of the two peaks is made the numerator in Eq. (1.4); the separation factor, consequently, always has values greater than or equal to 1.0. The separation factor is a measure of the selectivity of a chromatographic system. In thermodynamic terms it is related to the difference in free energy of the retention property responsible for the separation and is a term widely used in method development for defining systems with useful separation properties. To maintain a useful thermodynamic meaning the separation factor must be determined for fixed and constant experimental conditions, for example, constant temperature in gas chromatography and constant mobile phase composition in liquid chromatography. The separation factor is sometimes called the selectivity factor, selectivity or relative retention.

### 1.4.1 Influence of Mobile Phase Physical Properties

In pressure-driven systems a pressure gradient exists between the column inlet and outlet resulting in a change in volume-dependent terms over the length of the column

Table 1.2
Calculation of retention volumes in gas chromatography

Experimental data for calculation: retention time $t_R$ = 12.61 min; column hold-up time $t_M$ = 0.23 min; carrier gas flow rate at column outlet $F_a$ = 21.78 ml/min; column temperature $T_c$ = 121°C (394.2 K); ambient temperature $T_a$ = 23°C (296.2 K); ambient pressure $P_0$ = 754.5 mm Hg; column head pressure $P_G$ = 62.9 mm Hg; $P_w$ = vapor pressure of water at $T_a$ (available in handbooks of physical constants); and weight of liquid phase $W_L$ = 1.5115 g

• Calculation of the gas compressibility correction factor j
$j = 3 / 2 [(P^2 - 1) / (P^3 - 1)$ where P is the relative pressure (ratio of the column inlet pressure $P_i$ to the outlet pressure $P_0$). $P_i = P_G + P_0$ and $P = (62.9 + 754.5) / 754.5 = 1.0834$
$j = 3 / 2 (0.1737 / 0.2715) = 0.9596$

• Calculation of the carrier gas flow rate at the column temperature from the flow rate measured at the column outlet $F_0$. If measurements are made with a soap-film meter it is necessary to correct the flow rate for the difference between the dry gas (column) and water saturated gas (meter) measurements. $F_c$ is the corrected carrier gas flow rate.
$F_c = F_0 [T_c / T_a] [1 - (P_w / P_0)] = 21.78 [394.2 / 296.2] [1 - (21.068 / 754.5)] = 28.18$ ml / min.

• Column hold-up volume corresponding to the hold-up time is $V_M$ and after correction for compressibility of the carrier gas is the corrected hold-up volume $V_M°$. The latter is equivalent to the gas phase volume of the column at the average column pressure and temperature. $V_M = t_M F_c = (0.23) (28.18) = 6.48$ ml and $V_M° = j V_M = (0.9596) (6.48) = 6.22$ ml.

• The retention volume ($V_R$) is the volume of mobile phase entering the column between sample injection and the emergence of the peak maximum for the substance of interest. The corrected retention volume ($V_R°$) is the retention volume corrected for the compressibility of the carrier gas. $V_R = t_R F_c = (12.61) (28.18) = 355.3$ ml and $V_R° = j V_R = (0.9596) (355.3) = 341$ ml.

• The adjusted retention volume ($V_R'$) is the retention volume corresponding to the adjusted retention time and the net retention volume ($V_N$) is the adjusted retention volume corrected for the compressibility of the carrier gas. $V_R' = t_R' F_c = (12.61 - 0.23) (28.18) = 348.9$ ml and $V_N = j V_R' = (0.9596) (348.9) = 334.8$ ml.

• The specific retention volume ($V_g°$) is the net retention volume per gram of stationary phase (either liquid phase or solid adsorbent) at the column temperature. $V_g° = V_N / W_L = (334.8) / (1.5115) = 221.5$ ml/g. [the specific retention volume corrected to 0°C is $V_g = V_g°(273.2 / T_c)$]

that depends on the compressibility of the mobile phase. Mobile phase compressibility varies over a wide range with gases being the most compressible, liquids the least, and supercritical fluids in between. The mobile phase compressibility correction factor, j, allows the calculation of the average mobile phase velocity and solute retention volumes at the average column pressure from the experimentally measured inlet and outlet pressures in gas chromatography. The process is outlined for the example given in Table 1.2. The selection and correct use of the compressibility correction factor has generated some debate [14-19]. The gas compressibility correction factor can be specified for the column length or the solute residence time, and for ideal and non-ideal behavior of the carrier gas, all of which are different [19]. Typical usage, the conversion of volumes measured at the column outlet under ambient conditions into the corresponding volumes at the pressure averaged over the column length, assuming near ideal behavior for the carrier gas, is illustrated in Table 1.2. In gas-solid chromatography, the situation is somewhat different: gas and analyte molecules must compete for adsorption sites on the stationary phase and distribution constants are likely to be pressure dependent reflecting the influence of different gas density gradients over the column. Corrected retention volumes, therefore, are unlikely to be invariant of the column inlet pressure. The identity of the carrier gas should also play a more significant role in establishing the relative retention order in gas-solid chromatography, as generally observed [20].

Not only variations in the pressure at constant temperature influence column-to-column retention data: the role of the column hold-up volume as well as the mass of stationary phase present in the column is also important. The net retention volume calculated from the adjusted retention volume corrects for the column hold-up volume (see Table 1.2). The specific retention volume corrects for the different amount of stationary phase present in individual columns by referencing the net retention volume to unit mass of stationary phase. Further correction to a standard temperature of 0°C is discouraged [16-19]. Such calculations to a standard temperature significantly distort the actual relationship between the retention volumes measured at different temperatures. Specific retention volumes exhibit less variability between laboratories than other absolute measures of retention. They are not sufficiently accurate for solute identification purposes, however, owing to the accumulation of multiple experimental errors in their determination. Relative retention measurements, such as the retention index scale (section 2.4.4) are generally used for this purpose. The specific retention volume is commonly used in the determination of physicochemical properties by gas chromatography (see section 1.4.2).

It is normal practice to assume that the typical carrier gases used for gas chromatography are ideal. This allows volume corrections to be made using the ideal gas laws and for gas-solute interactions to be ignored in the interpretation of retention properties. For the most exact work, it may be necessary to allow for non-ideal behavior of the gas phase by applying a correction for solute-gas phase interactions [21,22]. For carrier gases that are insoluble in the stationary phase and at moderate column inlet pressures Eq. (1.5) is a reasonable approximation

$$\ln V_N = \ln V_N(0) + 0.75\,[(2B_{12} - V_1) / RT_c]\,[(P^4 - 1) / (P^3 - 1)]\,P_0 \qquad (1.5)$$

where $V_N(0)$ is the net retention volume at zero column pressure drop, $B_{12}$ the second interaction virial coefficient of the solute with the carrier gas, $V_1$ the solute molar volume at infinite dilution in the stationary phase (commonly replaced by the bulk molar volume), R the universal gas constant, P the relative pressure and $P_0$ the pressure at the column outlet (see Table 1.2 for definitions). Under normal operating conditions errors due to assuming ideality of the gas phase for simple carrier gases like hydrogen, helium and nitrogen are small, however, they increase with high solute concentrations, large column pressure drops, and low temperatures. Virial corrections are usually made only when it is desired to calculate exact thermodynamic constants from retention volume measurements. Alternatively, high-pressure gas chromatography can be used to calculate virial coefficients. The number of accurately determined virial coefficients is small and limits the general application of Eq. (1.5).

Liquids are far less compressible than gases and for the majority of applications the influence of the column pressure drop on the retention factor in liquid chromatography is ignored. Since the column pressure drop in normal and ultrahigh pressure liquid chromatography is relatively large this practice might be questionable in some cases [23-26]. The observed retention factor when calculated from retention time is an average value reflecting the retention factor gradient over the length of the column. The observed (average) retention factor has been shown to vary linearly with the inlet pressure in a solute-specific manner. The slope is usually shallow and unless retention factors are compared at large inlet pressure differences average retention factor values are nearly constant. However, it is now clear that retention factors are not invariant of pressure over the full range of inlet pressures used in modern liquid chromatography, and this has some implications for determining physicochemical properties by liquid chromatography but is less important for analysis.

Fluids are highly compressible and density gradients along the column associated with the column pressure drop result in significant retention factor changes as a function of local density. These changes are complex and usually modeled by empirical relationships (section 7.5).

### 1.4.2 Property Estimations

Gas chromatography is widely used to determine solution and adsorption thermodynamic properties [21,22,27-32]. Compared to classical static methods it has several advantages. Measurements can be made for impure samples, very small sample sizes are sufficient, and easy variation of temperature is provided. For the most exact measurements precise flow, pressure, and temperature control is needed that may require modification to a standard analytical gas chromatograph. The free energy, enthalpy, and entropy of mixing or solution, and infinite dilution solute activity coefficients can be determined from retention measurements made at infinite dilution (Henry's law region) in which the value of the activity coefficient (also the gas-liquid distribution constant) can be assumed to have a constant value. At infinite dilution the solute molecules are not sufficiently close to exert any mutual attractions, and the environment of each may be considered to consist entirely of solvent molecules. The activity coefficient and the specific retention volume are related by $V_g = (273.2\ R) / (M_2\gamma_1P_1{}^\circ)$ where $M_2$ is the molecular weight of the solvent, $\gamma_1$ the solute activity coefficient at infinite dilution, and $P_1{}^\circ$ the saturation vapor pressure of the pure solute at the given temperature. Ideally, activity coefficients calculated from the above relationship should be corrected for fugacity (solute-solute interactions), imperfect gas behavior, and interfacial adsorption. The first two corrections may introduce errors of ca. 1-5% in the value of the activity coefficient depending on the circumstances of the measurement; ignoring the importance of interfacial adsorption as a retention mechanism may make values for the activity coefficient completely meaningless (section 2.4.1). Typical infinite dilution activity coefficients for nonionic solvents, used in gas chromatography, have values in the range

0.3 to 50 [29,31]. Positive deviations from Raoult's law ($\gamma_1 > 1$) are common for the high-molecular-weight solvents generally used in gas chromatography. Activity coefficients much less than one indicate strong solute-stationary phase interactions.

The gas-liquid distribution constant ($K_L$), moles of solute per unit volume of liquid / moles of solute per unit volume of gas phase, is evaluated from the specific retention volume using the relationship $V_g = (273.2\ K_L) / (T_c \rho_c)$ where $\rho_c$ is the liquid phase density at the column temperature [32]. Alternatively, extrapolation of the net retention volume measured at several different phase loadings to an infinite stationary phase volume allows the gas-liquid distribution constant to be obtained independent of accompanying contributions from interfacial adsorption (section 2.4.1). The gas-liquid distribution constant can then be used to calculate values of the specific retention volume that are corrected for contributions to retention arising from interfacial adsorption. Also the partial molar Gibbs free energy of solution for a solute at infinite dilution ($\Delta G°$) in the stationary phase can be obtained directly from the gas-liquid distribution constant using $\Delta G° = -RT_c \ln K_L$. From the slope of a plot of log $V_g$ against the reciprocal of the column temperature over a narrow temperature range, 10-30 K, the enthalpy of solution is obtained. The entropy for the same process is obtained from a single value of the specific retention volume and the value of the enthalpy of solution calculated as just described [34-36].

Compared to gas chromatography liquid chromatography is used far less for physicochemical measurements [37,38]. Inadequate knowledge of the true composition of the stationary phase and the absence of quantitative models for the accurate description of retention are the principal reasons for this. A few exceptions are the determination of equilibrium constants that affect the form of a solute in the mobile phase (ion dissociation, complexation, confirmation, etc.) Also, indirect property determinations based on quantitative structure – activity relationships (QSAR) and quantitative structure – property relationships (QSPR) [39-43]. QSAR and QSPR relationships are based on the identification of an empirical correlation between a retention property in a chromatographic system, usually the retention factor and another (usually) equilibrium property of a chemical or biological system. Typical examples include the octanol-water distribution constant, the distribution of compounds across biological membranes, aquatic toxicity of organic compounds, the soil-water distribution constant, etc. These relationships are often, although not exclusively, of the form $\log P = a \log k + b$ where P is some equilibrium dependent property and a and b are empirical regression constants. Once the correlation equation is established using known values of log P further values of log P can be estimated from the correlation equation by measuring their chromatographic retention. This provides an inexpensive and rapid method for estimating properties that are difficult and expensive to determine by direct measurement.

### 1.4.3 Linear Free Energy Relationships

The free energy of transfer of a solute between two phases can be described as the linear sum of contributing processes delineated by a suitable model. For chromatographic and liquid-liquid distribution systems a cavity model provides a general approach for characterizing the contribution of solvent-solvent and solvent-solute interactions to equilibrium properties [44,45]. Firstly, a cavity of a suitable size to accommodate the solute is constructed in the solvent, with the solvent molecules in the same state as in the bulk solvent. The energy required for this process depends on the forces holding the solvent molecules together, and the solute's size. Cavity formation requires work and opposes solute transfer. In the second step the solvent molecules are reorganized into their equilibrium position round the solute. The free energy for this process is approximately zero and can be neglected. Although it should be pointed out, however, that the enthalpy and entropy of reorganization may be considerable – the free energy is effectively zero because of compensation, as in the melting of ice at 0°C. Finally, the solute is inserted into the cavity and various solute-solvent interactions are set up. For nonionic solutes these are identified as dispersion, induction, orientation, and hydrogen bonding. If two condensed phases are involved in the equilibrium then the free energy of transfer is equivalent to the difference in cavity formation and solute-solvent interactions in the two phases. For transfer from an ideal gas phase to a solvent at infinite dilution the free energy of transfer is equal to the difference in free energy of cavity formation in the solvent and the strength of solute-solvent interactions.

To move from a qualitative to a quantitative picture the individual free energy contributions to the solvation process identified above must be delineated in a quantitative form. Within the framework of a linear free energy relationship the contributions of individual intermolecular interactions are represented as the sum of product terms made up of solute factors (descriptors) and complementary solvent factors (system constants). Thus a solute has a certain capability for a defined intermolecular interaction and its contribution to the solution free energy is the product of the capability of the solute and solvent for that interaction. Kamlet, Taft and their co-workers [44,46] developed one of the earliest general approaches to the quantitative characterization of solute-solvent interactions based on solvatochromism. Solvatochromic parameters were defined by the influence of environment (solvent effects) on the absorption spectra of select compounds and normalized to provide roughly equivalent scales. This method has been widely used to determine the dipolarity/polarizability ($\pi^*$), the hydrogen-bond acidity ($\alpha$) and hydrogen-bond basicity ($\beta$) of common solvents [47]. Kamlet, Taft and their co-workers extended their solvatochromic parameters to solute effects, assuming that the solvent parameters could be taken as an estimate of solute properties. This is at best a rough approximation. In a bulk solvent, each molecule is surrounded by molecules like itself, while as a solute it is surrounded by solvent molecules that are different to it. Compounds, such as alcohols, that are highly associated as solvents are expected to behave differently as monomeric solute molecules. However, there are also fundamental limitations to this approach. The solvatochromic parameters are related to spectroscopic energy differences, that is the influence of solvent effects on the ground and excited states of the selected indicator compounds, which are not free energy processes *per se*. Secondly, although some parameter estimate rules have been developed, there is no protocol for the determination of the Kamlet-Taft parameters for additional (especially solid) compounds. In order to construct a correlation equation that has a sound physical interpretation, it is necessary that the various descriptors should be related to Gibbs free energy. Descriptors meeting this requirement were developed by Abraham and co-workers [45,48-51] and are to be preferred to the solvatochromic parameters for chromatographic retention studies and for wider application to solute properties that can be characterized by a distribution constant. Before describing Abraham's solvation parameter model, it is necessary to reiterate that the solute descriptors for the solvation parameter and solvatochromic models are not the same, although they are often mistaken or misused as such in the contemporary literature. The solvation parameter model is also unrelated to the solubility parameter model.

The solvation parameter model for distribution between two condensed phases, Eq. (1.6) or (1.6a), and transfer from the gas phase to a solvent, Eq. (1.7) or (1.7a), are set out below in the form generally used in chromatography.

$$\log SP = c + mV_X + rR_2 + s\pi_2^H + a\sum\alpha_2^H + b\sum\beta_2^H \tag{1.6}$$

$$\log SP = c + vV + eE + sS + aA + bB \tag{1.6a}$$

$$\log SP = c + rR_2 + s\pi_2^H + a\Sigma\alpha_2^H + b\Sigma\beta_2^H + l\log L^{16} \tag{1.7}$$

$$\log SP = c + eE + sS + aA + bB + lL \tag{1.7a}$$

Eqs. (1.6) and (1.6a) and (1.7) and (1.7a) are identical but written with different symbols. Eqs. (1.6) and (1.7) have been commonly used in the literature following Abraham's description of the solvation parameter model. Recently, Abraham has suggested replacement of these equations with (1.6a) and (1.7a) to simplify representation of the model [52,53]. It is likely that Eqs. (1.6a) and (1.7a) will replace Eqs. (1.6) and (1.7) as the general representation of the solvation parameter model in the future.

SP is some free energy related solute property such as a distribution constant, retention factor, specific retention volume, relative adjusted retention time, or retention index value. Although when retention index values are used the system constants (lowercase letters in italics) will be different from models obtained with the other dependent variables. Retention index values, therefore, should not be used to determine system properties but can be used to estimate descriptor values. The remainder of the equations is made up of product terms called system constants (*r*, *s*, *a*, *b*, *l*, *m*) and solute descriptors ($R_2$, $\pi_2^H$, $\Sigma\alpha_2^H$, $\Sigma\beta_2^H$, $\log L^{16}$, $V_X$). Each product term represents a contribution from a defined intermolecular interaction to the solute property. The contribution from cavity formation and dispersion interactions are strongly correlated with solute size and cannot be separated if a volume term, such as the characteristic volume [$V_X$ in Eq. (1.6) or V in Eq. (1.6a)] is used as a descriptor. The transfer of a solute between two condensed phases will occur with little change in the contribution from dispersion interactions and the absence of a specific term in Eq. (1.6) to represent dispersion interactions is not a serious problem. For transfer of a solute from the gas phase to a condensed phase this

Table 1.3
Calculation of solute descriptor values for the solvation parameter model

• Calculation of McGowan's characteristic volume, $V_X$ (or V), for toluene
Atomic volumes: C = 16.35, H = 8.71, N = 14.39, O = 12.43, F = 10.48, Si = 26.83, P = 24.87, S = 22.91, Cl = 20.95, B = 18.23, Br = 26.21, I = 34.53. Subtract 6.56 for each bond of any type.
Toluene = 7 carbon atoms + 8 hydrogen atoms – 15 bonds = 114.45 + 69.68 – 98.40 = 85.73 in $cm^3.mol^{-1}$. After scaling $V_X$ = 0.857 in $cm^3.mol^{-1}/100$.

• Calculation of the excess molar refraction, $R_2$ (or E), for toluene using Eq. (1.8). The refractive index for toluene ($\eta$) at 20°C (sodium D-line) = 1.496
$R_2 = 8.57 (0.292) + 0.5255 - 2.832 (0.857) = 0.601$ in $cm^3.mol^{-1}/10$.

• Estimation of solute descriptors for 2,6-dimethoxyphenol from liquid-liquid distribution constants. $V_X$ and $R_2$ were calculated as above giving 1.1743 and 0.840, respectively. Other solute descriptors were obtained as the best-fit values from the distribution systems given below

| Distribution system | log K(calc.) | log K(exp.) | Best-fit values | | |
|---|---|---|---|---|---|
| | | | $\pi_2^H$ | $\Sigma\alpha_2^H$ | $\Sigma\beta_2^H$ |
| Water-octanol | 1.10 | 1.15 | | | |
| Water-ether | 0.79 | 0.74 | 1.41 | 0.13 | 0.71 |
| Water-olive oil | 0.56 | 0.57 | | | |
| Water-hexadecane | -0.35 | -0.36 | | | |
| Water-cyclohexane | -0.15 | -0.15 | | | |

is no longer the case and the solvation equation must be set up to account for the contribution of dispersion interactions to the free energy of solute transfer. Abraham handled this problem by defining a second descriptor for the contribution of cavity formation and dispersion interactions [log $L^{16}$ in Eq. (1.7) or L in Eq. (1.7a)]. This term includes not only solute-solvent dispersion interactions, but also the cavity effect making the $V_X$ term in Eq. (1.6) redundant. For general applications Eq. (1.6) is the form of the model suitable for characterizing chromatographic retention in systems with two condensed phases, such as liquid and micellar electrokinetic chromatography. Eq. (1.7) is suitable for characterizing retention in gas chromatography, and more generally in two phase systems were one component is a gas.

The solute descriptors used in Eq. (1.6) and (1.7) must be free energy related properties to correlate with chromatographic retention. It is also important that the solute descriptors are accessible for a wide range of compounds by either calculation or simple experimental techniques, otherwise the models would lack practical utility. McGowan's characteristic volume, $V_X$ or V in units of $cm^3.mol^{-1}$ / 100, can be calculated for any molecule whose structure is known by simple summation rules, Table 1.3 [49,54]. Each atom has a defined characteristic volume and the molecular volume is the sum of all atomic volumes less 6.56 $cm^3.mol^{-1}$ for each bond, no matter whether single, double or triple. For complex molecules the number of bonds, B, is easily calculated from the algorithm $B = N - 1 + R$ where N is the total number of atoms, and R is the number of rings. Log $L^{16}$ or L is the solute gas-liquid distribution constant (also referred to as the Ostwald solubility coefficient) on hexadecane at 298 K. For volatile solutes it can be determined directly [50]. For all compounds of low volatility, it is determined by back calculation from gas chromatographic retention measurements on nonpolar stationary phases at any convenient temperature [56-60]. Suitable stationary phases are those for which the system constants $a \approx b \approx s \approx 0$ in Eq. (1.7).

The solute excess molar refraction, $R_2$ or E, models polarizability contributions from n- and $\pi$-electrons. The solute molar refraction is too closely related to solute size to be used in the same correlation equation as $V_X$. To avoid correlation between the molar refraction and $V_X$, an excess molar refraction, $R_2$, was defined as the molar refraction for the given solute, less the molar refraction for an n-alkane of the same characteristic volume [61,62]. The excess molar refraction is simply calculated from the refractive index of the solute at 20°C for the sodium D-line, $\eta$, as indicated by Eq. (1.8)

$$R_2 = 10V_X[(\eta^2 - 1) / (\eta^2 + 2)] - 2.832\, V_X + 0.526 \qquad (1.8)$$

The units used for $V_X$ in Eq. (1.8) are $cm^3.mol^{-1}/100$, and therefore $R_2$ is given in $cm^3.mol^{-1}$ / 10. The use of Eq. (1.8) to calculate the excess molar refraction is straightforward for liquids but even for solids refractive index values are easily estimated using available software for molecular property estimations. In addition, $R_2$, like the molar refraction, is almost an additive quantity, and values for solids can be estimated through addition of fragments with known $R_2$ values [45,63-65].

In developing the solvation parameter model Abraham and coworkers commenced the process by defining descriptors for solute hydrogen-bond acidity ($\alpha_2^H$ )and solute hydrogen-bond basicity ($\beta_2^H$ ). The superscript (H) indicates the origin of the scale and the subscript (2) that the descriptors are solute properties. Initially these solute descriptors were determined from 1:1 complexation constants measured in an inert solvent [66,67]. These studies also led to scales that had a zero origin. A problem still remained, however, when these descriptors were used to characterize distribution processes. The influence of solute structure on the distribution process will be a consequence of hydrogen bonding of the solute to any surrounding solvent molecules, not just to one. What are needed are scales of "summation" or "overall" hydrogen bonding that refer to the propensity of a solute to interact with a large excess of solvent molecules. These hydrogen-bond descriptors are denoted as $\sum \alpha_2^H$ and $\sum \beta_2^H$ to distinguish them from the 1:1 descriptors. New values of the effective hydrogen bonding solute descriptors are now determined in conjunction with other solute descriptors using liquid-liquid distribution and chromatographic measurements [49,68,69]. A minor complication is that certain solutes (sulfoxides, anilines, pyridines) show variable hydrogen-bond basicity in distribution systems where the organic phase absorbs appreciable amounts of water [68]. A new solute descriptor $\Sigma\beta_2^0$ was defined for these solutes and should be used in octanol-water distribution systems, for example, and for reversed-phase and micellar electrokinetic chromatography. For the same solutes $\sum \beta_2^H$ should be used for all other applications and always for gas chromatography. Except for the solute types indicated above, the two hydrogen-bond basicity scales are identical. It should also be noted that the scales of hydrogen-bond acidity and basicity are generally unrelated to proton transfer acidity and basicity expressed by the $pK_a$ scale.

It would be useful to have descriptors that were related to the propensity of a solute to engage in dipole-dipole and induced dipole-dipole interactions. In the event, it proved impossible to separate out descriptors for the two types of interactions, and Abraham and coworkers [61] constructed a solute descriptor for dipolarity/polarizability, $\pi_2^H$ or S, combining the two interactions. The dipolarity/polarizability descriptor was initially determined through gas chromatographic measurements on polar stationary phases [56,57,70], but is now more commonly determined in combination with the hydrogen-bonding solute descriptors from liquid-liquid distribution constants and chromatographic measurements [45,50,69].

Solute descriptors are available for about 4,000 compounds, with some large compilations reported [42,43,49,58,68]. For additional compounds estimation methods are available using fragment constants [46,50,63-65]. An early version of a software program to estimate solute descriptors from structure has appeared [71]. In all other cases it is possible to calculate $V_X$ and $R_2$ and determine the other descriptors from experimental distribution constants and chromatographic measurements. If data is available for a particular solute in three systems with significantly different system constants then $\pi_2^H$, $\Sigma\alpha_2^H$, $\Sigma\beta_2^H$ can be determined as the solution to three simultaneous equations. If the number of equations is larger than the number of descriptors to be determined, the descriptor values that give the best-fit solution (i.e., the smallest standard deviation in the observed and calculated log SP values) are taken [69].

The system constants in Eqs. (1.6) and (1.7) are obtained by multiple linear regression analysis for a number of solute property determinations for solutes with known descriptors. The solutes used should be sufficient in number and variety to establish the statistical and chemical validity of the model [72-74]. In particular, there should be an absence of significant cross-correlation among the descriptors, clustering of either descriptor or dependent variable values should be avoided, and an exhaustive fit should be obtained. Table 1.4 illustrates part of a typical output. The overall correlation coefficient, standard error in the estimate, Fischer F-statistic, and the standard deviation in the individual system constants are used to judge whether the results are statistically sound. An exhaustive fit is obtained when small groups of solutes selected at random can be deleted from the model with minimal change in the system constants.

The system constants are more than regression constants and contain important chemical information about the system. Consequently, not only must the system constants be statistically sound but they must make chemical sense as well. The system constants reflect the difference in solute interactions in the two phases, except for gas chromatography, where the system constants reflect stationary phase properties alone. The *r* (or *e*) system constant indicates the tendency of the phases to interact with solutes through π- and n-electron pairs; the *s* system constant for the tendency of the phases to interact with solutes through dipole-type interactions; the *a* system constant denotes the difference in hydrogen-bond basicity between phases (because acidic solutes will interact with a basic phase); and the *b* system constant is a measure of the difference in hydrogen-bond acidity of the phases (because basic solutes will interact with an acidic phase). The *l* system constant is a measure of the energy required for cavity formation and the strength of dispersion interactions in gas chromatography. The *m* (or *v*) system constant is a measure of the difference in cavity formation in the two condensed phases

Table 1.4
An example of part of the output for fitting the solvation parameter model to a reversed-phase chromatographic system by multiple linear regression analysis

| Solute | Descriptors | | | | | log k | |
|---|---|---|---|---|---|---|---|
| | $V_X$ | $R_2$ | $\pi_2^H$ | $\Sigma\alpha_2^H$ | $\Sigma\beta_2^0$ | Experimental | Predicted |
| Phenol | 0.775 | 0.805 | 0.89 | 0.60 | 0.31 | -0.306 | -0.273 |
| Benzyl Alcohol | 0.916 | 0.803 | 0.87 | 0.33 | 0.56 | -0.268 | -0.252 |
| Aniline | 0.816 | 0.955 | 0.96 | 0.26 | 0.50 | -0.386 | -0.380 |
| Toluene | 0.857 | 0.601 | 0.52 | 0 | 0.14 | 0.553 | 0.524 |
| Ethylbenzene | 0.998 | 0.613 | 0.51 | 0 | 0.15 | 0.997 | 0.937 |
| Naphthalene | 1.085 | 1.340 | 0.92 | 0 | 0.20 | 1.185 | 1.256 |
| Benzaldehyde | 0.873 | 0.820 | 1.00 | 0 | 0.39 | -0.017 | -0.011 |
| Nitrobenzene | 0.890 | 0.871 | 1.11 | 0 | 0.28 | 0.143 | 0.225 |
| Chlorobenzene | 0.838 | 0.718 | 0.65 | 0 | 0.07 | 0.618 | 0.597 |
| Acetophenone | 1.014 | 0.818 | 1.00 | 0 | 0.49 | 0.275 | 0.236 |

| *Cross-correlation matrix ($r^2$)* | | | | | | *Model Statistics* | *System constants* |
|---|---|---|---|---|---|---|---|
| $V_X$ | 1.00 | | | | | | $m$ = 2.99 (0.07) |
| $R_2$ | 0.19 | 1.00 | | | | R = 0.994 | $r$ = 0.46 (0.05) |
| $\pi_2^H$ | 0.00 | 0.33 | 1.00 | | | SE = 0.07 | $s$ = -0.44 (0.05) |
| $\Sigma\alpha_2^H$ | 0.29 | 0.22 | 0.03 | 1.00 | | F = 570 | $a$ = -0.30 (0.05) |
| $\Sigma\beta_2^0$ | 0.00 | 0.03 | 0.42 | 0.17 | 1.00 | n = 40 | $b$ = -1.88 (0.08) |
| | | | | | | | $c$ = -1.82 (0.07) |

together with any residual dispersion interactions that are not self-canceling. For the reversed-phase chromatographic system summarized in Table 1.4 the factors leading to retention (increase in log k) are the favorable cohesion properties of the stationary phase and electron lone pair interactions ($m$ and $r$ system constants have a positive sign). All polar interactions favor solubility in the mobile phase ($s$, $a$,and $b$ system constants have a negative sign) and result in reduced retention. From the relative magnitude of the system constants the most important solute property resulting in increased retention is its size and the most important factor reducing retention is its hydrogen-bond basicity. These are general properties of reversed-phase chromatographic systems. Applications of the solvation parameter model for material characterization, retention prediction, and method development in gas chromatography (section 2.3.5), reversed-phase liquid chromatography (section 4.3.1.3), thin-layer chromatography (section 6.7.2), micellar electrokinetic chromatography (section 8.3.3) and solid-phase extraction [75,76] are discussed elsewhere.

### 1.4.4 Exothermodynamic Relationships

A number of general relationships are used in chromatography to relate retention to solute or experimental variables. Apart from the influence of temperature these are not formal thermodynamic relationships but nevertheless are very useful for predicting retention properties and establishing retention mechanisms. They can be formulated as linear free energy relationships on the assumption that the free energy of solute transfer from the mobile to the stationary phase is an additive property. In which case the free energy of transfer can be taken as the sum of the free energy for each structural element of the solute. By way of example, for compounds belonging to a homologous series, such as $CH_3(CH_2)_nOH$, the total free energy of transfer can be decomposed into $\Delta G°(CH_3) + n\Delta G°(CH_2) + \Delta G°(OH)$. If we further assume that the difference between the methylene group and methyl group can be ignored then linear plots are expected for a plot of the logarithm of the retention property against carbon number (n or n + 1 in this case). This is generally what is observed in gas chromatography at a constant temperature and liquid chromatography at a constant mobile phase composition when $n \geq 3$ [77-80]. The logarithm of the chromatographic retention property can be any of the following; adjusted retention time, adjusted retention volume, specific retention volume, retention factor, or distribution constant. In the case of gas-liquid chromatography the saturation vapor pressure, boiling point or refractive index can be substituted for the carbon number, since these properties are often highly correlated with the carbon number for members of a homologous series. These relationships are sometimes referred to as Martin's rule, after the pioneer of partition chromatography (A. J. P. Martin) who anticipated and demonstrated their existence. From this concept Martin also explained the remarkable power of chromatography to separate substances that differ only slightly in their structure. For two compounds that differ in a structural element the difference in free energy of transfer (separation) is proportional to the corresponding free energy change for that structural element but not for the rest of the molecule. This explains why proteins that differ only in a single amino acid can be separated. As an extension of this idea functional group contributions for a large number of substituent groups on a range of parent structures have been derived to predict retention of simple substituted compounds from parent structures in reversed-phase liquid chromatography [77,78] and gas-liquid chromatography [40]. For polyfunctional compounds this approach has met with limited success because the additivity principle fails to account for intramolecular interactions between substituent groups and steric factors which influence solute-solvent interactions in a solute characteristic manner.

A similar linear logarithmic relationship, known as a van't Hoff plot, usually exists between adjusted retention data and the reciprocal of column temperature in gas, liquid (constant composition) and supercritical fluid (constant density) chromatography. The effect of temperature on retention is based on the Gibbs-Helmholtz equation and has a sound thermodynamic basis, Eq. (1.9)

$$\ln k = - (\Delta H° / RT) + (\Delta S° / R) + \ln \beta \qquad (1.9)$$

where $\Delta H°$ is the standard enthalpy and $\Delta S°$ the standard entropy of transfer of the solute from the mobile to stationary phase, R the gas constant, T temperature (K) and $\beta$ the phase ratio. The plot of ln k against 1 / T will be linear provided the change in heat capacity for the transfer is zero (i.e., $\Delta H°$, $\Delta S°$ and $\beta$ are temperature invariant). Nonlinear plots with sharp discontinuities in their slopes are observed for mixed retention mechanisms or change in solute or stationary phase conformation that influences sorption [79,81,82]. Linear van't Hoff plots can be used to determine enthalpies (from the slope) and entropies of transfer (from the intercept) although the latter can be difficult to determine owing to the non-trivial nature of calculating the phase ratio, particularly for liquid and supercritical fluid chromatography.

The influence of temperature on resolution is not as straightforward to predict as that of retention. Changing temperature affects thermodynamic properties (retention) and kinetic properties (peak shapes) at the same time [81-83]. Higher temperatures in gas chromatography decrease the cohesive properties and capacity of stationary phases for polar interactions resulting in a characteristic solute dependent reduction in retention [50,84,85]. For mixtures with a narrow range of retention properties there is usually an optimum temperature at which peak separations are maximized. In practice higher temperatures than this may be preferred if an adequate peak separation is maintained, more favorable peak shapes result, and faster separations are obtained. This is because temperature also influences diffusion coefficients in the mobile and stationary phases and the viscosity of the carrier gas. In general, the operating pressure will not be the limiting factor in gas chromatography and higher inlet pressures can be used to maintain an optimum mobile phase velocity to offset the increase in the carrier gas viscosity at higher temperatures. Higher temperatures, as well, result in improved mass transfer properties and narrower peaks counteracting any decrease in peak spacing. The net result is that higher temperatures can result in changes in resolution that can be for the better or worse.

The vast majority of liquid chromatographic separations are carried out at ambient temperature for convenience and because ambient temperature provides reasonable column efficiency for low molecular weight solutes. Elevated temperatures reduce the viscosity of the mobile phase and enhance solute diffusion [81,86,87]. This is expected to result in lower operating pressures, improved mass transfer properties but increased longitudinal diffusion (these properties are discussed in section 1.5.2). In terms of peak shapes this can be advantageous or disadvantageous depending on the mobile phase velocity and whether longitudinal diffusion or mass transfer properties dominate. Higher temperatures will also change the column selectivity so that predicting changes in resolution as a function of temperature can be difficult. The simultaneous optimization of temperature and mobile phase composition is considered desirable for method development in liquid chromatography, although the possibility of exploiting temperature as an optimization variable is often ignored [87-89]. Temperature variation and composition variation show similar trends in retention, but within the easily accessible range for both variables the capacity to change retention is much greater for composition variation [89,90]. The predominant influence of higher temperatures in reversed-phase liquid chromatography is to decrease retention by a reduction in the difference in cohesive energy between the mobile and stationary phases and to decrease the hydrogen-bond acidity of the mobile phase relative to the stationary phase, Figure 1.3 [89]. Changes in other polar interactions are less significant. Temperature variation in reversed-phase liquid chromatography, therefore, will have the largest effect on peak spacing of compounds that differ in size and hydrogen-bond basicity. The

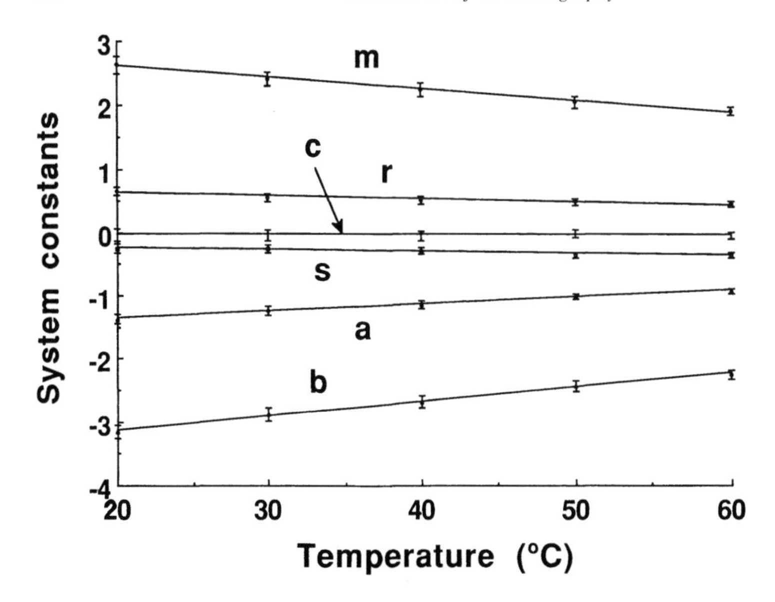


Figure 1.3. Variation of the system constants of the solvation parameter model (section 1.4.3) with temperature for 37 % (v/v) propan-2-ol in water on the porous polymer PLRP-S stationary phase. The *m* constant reflects the difference in cohesion and dispersive interactions, *r* constant loan-pair electron interactions, *s* constant dipole-type interactions, *a* constant hydrogen-bond basicity and *b* constant hydrogen-bond acidity between the mobile and stationary phases. (From ref. [89]; ©The Royal Society of Chemistry).

use of very high temperatures in liquid chromatography is quite recent, for example, pressurized water up to 200°C [90], the separation of proteins on nonporous sorbents at up to 120°C [91], and the separation of varied compounds on polymer encapsulated sorbents at up to 200°C [92]. Here the main interest has been to exploit pure water as a mobile phase or to obtain fast separations by increasing the separation efficiency at high flow rates. These extreme temperatures are manageable from an instrumental perspective but careful stationary phase selection is important because many chemically bonded phases degrade rapidly at high temperatures. Subambient temperatures are generally reserved for the separation of labile solutes where the inefficient column operation, long separation times, and high column pressure drops are justified by the need to maximize the peak separation of enantiomers or protect the solute from decomposition, denaturation, or conformational changes [83,93,94].

Plots of the standard entropy, or better standard free energy, against the standard enthalpy referred to as enthalpy-entropy compensation plots, are a useful tool for establishing the similarity of the retention mechanism for different solutes [25,79,83,95,96]. When enthalpy-entropy compensation occurs the plots are linear and the slope is called the compensation temperature. All compounds have the same retention at the compensation temperature, although their temperature dependence may differ. This is most likely to occur for processes governed by a single retention mechanism. All related processes that have the same compensation temperature proceed via the same mechanism. Some care is needed in the interpretation of these plots though. It is entirely possible to observe adventitious correlations between the standard enthalpy and entropy from spurious statistical effects that arise from the least squares fitting of the data [95-97]. This can happen when the entropy and enthalpy are both obtained from the same van't Hoff plot. When the enthalpy-entropy compensation is real the plot of the retention factor (ln k) against the reciprocal of temperature for all solutes must show a single intersection point. In addition the enthalpy-entropy compensation can be considered real only when the analysis of variance (ANOVA) indicates that the variance explained by the plot exceeds the random experimental variance according to an appropriate F-test. When a highly variegated set of solutes is chosen and enthalpy-entropy compensation is observed then one can be reasonably certain the retention process is dominated by a single factor. Enthalpy-entropy compensation is invariably observed for members of a homologous series since these have virtually an identical capacity for polar interactions and differ only by size. Besides temperature, enthalpy-entropy compensation can be observed for other system parameters. For example, organic modifier concentration in reversed-phase liquid chromatography, molal salt concentration in hydrophobic interaction chromatography, or a stationary phase property such as the carbon number of different bonded alkyl chains [79].

### 1.4.5 General Elution Problem

Separations employing constant temperature in gas chromatography, constant mobile phase composition in liquid chromatography, and constant density in supercritical fluid chromatography fail to provide useful results for samples containing components with a wide range of distribution constants. In gas chromatography, for example, there is an approximate exponential relationship between retention time and solute boiling point at a constant (isothermal) column temperature. Consequently, it is impossible to establish a suitable compromise temperature for the separation of mixtures with a boiling point range exceeding about 100°C. In liquid and supercritical fluid chromatography similar problems arise when the affinity of the solute for the stationary phase is sufficient to preclude convenient elution at a constant mobile phase composition or density, respectively. This is generically referred to as "the general elution problem" and is characterized by long separation times, poor separations of early eluting peaks and poor detectability of late eluting peaks due to band broadening. The general solution to this problem is the use of programmed separation modes. For gas chromatography the useful program modes are temperature and flow programming (section 2.4.3); for liquid chromatography mobile phase composition, flow and temperature programming (section 4.4.6); and for supercritical fluid chromatography density, composition and temperature programming (section 7.5.1). The programmed modes provide a complete separation of mixtures with a wide range of retention properties in a reasonable time. Neither constant nor programmed modes are superior to one another. They are complementary with the properties of the sample deciding which approach is adopted.

Temperature programming is the most popular programmed separation mode in gas chromatography. Stationary phases of high thermal stability allow wide temperature ranges to be used and temperature is easily adjusted and controlled using the forced air circulation ovens in general use with modern instruments. Flow programming is easily achieved with instruments fitted with electronic pressure control but is limited by the narrow pressure range usually available. It can be used to separate thermally labile compounds at a lower temperature than required for temperature programmed separations. On the other hand, flow programming results in a loss of efficiency for late eluting peaks and presents difficulties in calibrating flow sensitive detectors. Composition programming is the common programmed separation mode in liquid chromatography. Temperature and flow programming are not generally used except in special circumstances. The high flow resistance of packed columns and limited operating pressures results in a narrow range of flow rates and therefore retention variation compared with composition variation. In addition, volumetric dilution of later eluting peaks reduces sample detectability with concentration-sensitive detectors. Radial temperature gradients in normal bore columns render temperature programming impractical, causing a loss of efficiency and the formation of asymmetric peaks [98]. The smaller radius and reduced mass of microcolumns makes them more amenable to temperature programming techniques where this method of reducing separation times becomes more interesting on account of the greater difficulty of constructing gradient forming devices with very low mixing volumes [98-102]. In supercritical fluid chromatography density and composition programmed modes are both popular. When a single fluid is used as mobile phase then density programming is used. For fluids mixed with an organic solvent composition programming is more common although density programming may still be used, particularly at low solvent compositions. Simultaneous temperature programs with density and composition programs are sometimes applied to improve band spacing in the separation.

## 1.5 BAND BROADENING

If the sample is introduced as a sharp rectangular pulse into a column the individual separated sample components when they leave the column are broadened about their characteristic retention value in proportion to the time each component remained in the column. This characteristic change in the appearance of bands in the chromatogram results from kinetic factors referred to in total as band broadening. Zone is sometimes used for band, and dispersion or spreading for broadening, resulting in a number of names for the same process. For consistency we will call the process band broadening. The extent of band broadening determines the chromatographic efficiency, conventionally expressed as either the number of theoretical plates or simply the plate number (N), or the height equivalent to a theoretical plate (HETP) or simply plate height (H). If the column is assumed to function as a Gaussian operator then the column efficiency is readily expressed in terms of the peak retention time ($t_R$) and

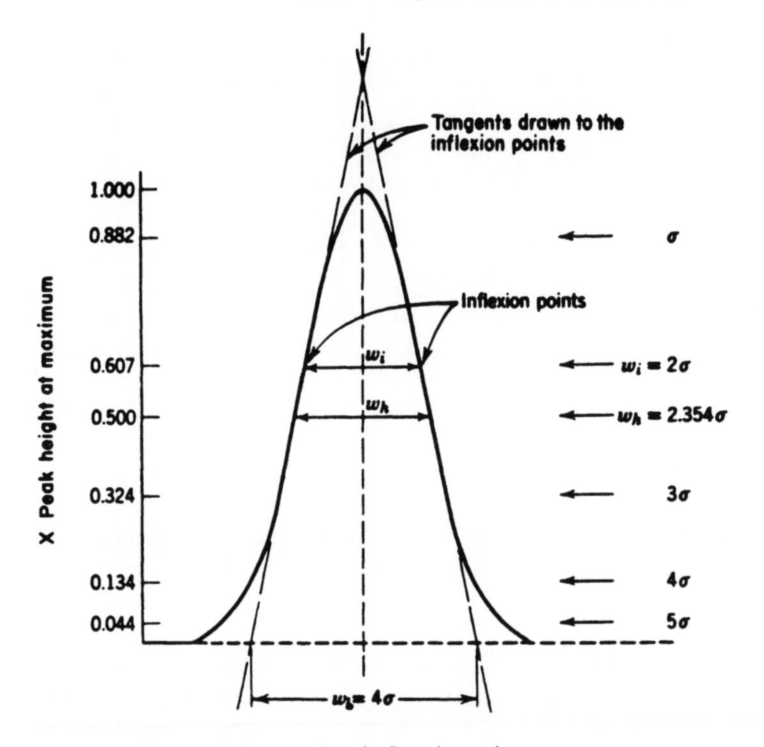


Figure 1.4. Characteristic properties of a Gaussian peak.

band variance in time units ($σ_t$) according to $N = (t_R / σ_t)^2$. In practice, various peak width measurements are frequently used based on the properties of a Gaussian peak, Figure 1.4, and Eq. (1.10)

$$N = a(t_R / w)^2 \qquad (1.10)$$

where $w_i$ is the peak width at the inflection point when a = 4, $w_h$ the peak width at half height when a = 5.54, and $w_b$ the peak width at the base when a = 16. Alternatively the ratio of the peak height to the area of a Gaussian peak can be used to define N

$$N = 2π(t_R h / A)^2 \qquad (1.11)$$

where h is the peak height and A the peak area. The plate height, H, is given by the ratio of the column length (L) to the column plate count by H = L / N. Column efficiency can also be measured as the effective plate number ($N_{eff}$) by substituting the adjusted retention time, $t_R$' for the retention time in Eqs. (1.9) and (1.10). The effective plate number is considered more fundamentally significant than the plate number since it measures only the band broadening that occurs during the time the solute interacts with the stationary phase. The two measures of column efficiency are related by $N_{eff}$ = N [k /

$(1 + k)]^2$ where k is the retention factor. For a weakly retained solute, for example $k = 1$, $N_{eff}$ will be only 25% of the value of N; however, for well-retained solutes, $k > 10$, $N_{eff}$ and N will be approximately the same. For useful column comparisons $N_{eff}$ and N should be determined for well-retained solutes; at low k values N will be speciously high and misrepresent the actual separation performance that can be obtained from a particular column in normal use. Also for comparative purposes, it is general practice to normalize the value of $N_{eff}$ and N on a per meter of column length basis. For many of the relationships discussed in this chapter, N and $N_{eff}$ can be used interchangeably.

The terms plate number and plate height have their origin in the plate model of the chromatographic process [103-108]. This model, originally proposed by Martin and Synge in 1941 [109], was an extension of distillation and countercurrent liquid-liquid distribution theory to the new technique of partition chromatography. It is only of historic interest now, having been replaced by more realistic rate theory models. The plate model, however, has contributed significantly to the terms used to describe band broadening in chromatography and is briefly discussed for that reason. The plate model assumes that the column can be visualized as being divided into a number of volume elements or imaginary sections called plates. At each plate the partitioning of the solute between the mobile and stationary phase is rapid and equilibrium is reached before the solute moves onto the next plate. The solute distribution constant is the same in all plates and is independent of the solute concentration. The mobile phase flow is assumed to occur in a discontinuous manner between plates and diffusion of the solute in the axial direction is negligible (or confined to the volume element of the plate occupied by the solute). The main problem with the plate model is that it bears little physical resemblance to the chromatographic process. Axial diffusion is a significant source of band broadening in chromatography, the distribution constant is independent of concentration only over a narrow concentration range, and, quite obviously, the assumption that flow occurs in a discontinuous manner is false. In the practical sense its largest shortcoming is that it fails to relate the band broadening process to the experimental parameters (e.g., particle size, mobile phase velocity, etc.) that are open to manipulation by the investigator. Nevertheless, the measured quantities N and H are useful parameters for characterizing chromatographic efficiency and are not limited by any of the deficiencies in the plate model itself. More realistic models developed from rate theory enable a similar expression for the plate number to be derived. The general connection is that all these models result in a Gaussian-like distribution for the band broadening process, but from rate theory we can relate band broadening to the experimental variables, and in that way establish an experimental basis for kinetic optimization of the chromatographic process.

### 1.5.1 Flow Through Porous Media

For an understanding of band broadening in chromatographic systems, the linear velocity of the mobile phase is more important than the column volumetric flow rate. Complications in identifying a suitable velocity arise for compressible mobile phases with pressure-driven systems. In this case the local velocity at any position in the column will depend on the flow resistance of the column and the ratio of the column inlet to outlet pressure. An average linear velocity, u, is always available as the ratio of the column length to the retention time of an unretained solute ($L / t_M$). For open tubular columns this definition is unambiguous. For packed columns the measured value will depend on the ability of the unretained solute to probe the pore volume of the stationary phase. Two extreme values are possible for an unretained solute with total access to the pore volume and an unretained solute that is fully excluded from the pore volume. For porous stationary phases the mobile phase trapped within the pores is generally stagnant and for kinetic optimization the mobile phase velocity through the interparticle volume, the interparticle velocity, $u_e$, is probably more fundamentally significant [110,111].

Since many chromatographic experiments are performed with the column exit at ambient conditions, and many solute properties are known for these conditions, the experimentally accessible outlet velocity, $u_o$, can be useful. The mobile phase velocity and flow rate in an open tubular column are simply related by $u_o = F_c / A_c$ where $F_c$ is the fully corrected column volumetric flow rate and $A_c$ the column cross-sectional area available to the mobile phase. In a packed bed the flow of mobile phase occurs predominantly through the interparticle spaces and the mobile phase velocity at the column outlet is thus described by the equation $u_o = F_c / \pi r_c^2 \varepsilon_u$ where $r_c$ is the column radius and $\varepsilon_u$ the interparticle porosity. For well-packed columns $\varepsilon_u$ is about 0.4. For gas chromatography under normal operating conditions with a modest pressure drop the average and outlet velocity are simply related by $u = ju_o$, where j is the gas compressibility correction factor (see section 1.4.1). At high-pressure drops, as could exist in fast gas chromatography, the average carrier gas velocity becomes proportional to the square root of the outlet velocity [112]. For supercritical fluid chromatography the correction is more complicated, in part, because fluids behave non-ideally [113].

The mobile phase flow profile and changes in local velocity depend on the driving force used to maintain bulk flow through the separation system. These driving forces can be identified as capillary, pneumatic or electroosmotic forces. Capillary forces are responsible for the transport of mobile phase in conventional column liquid chromatography and planar chromatography. These forces are generally too weak to provide either an optimum or constant mobile phase velocity for separations using small particle sorbents. They are treated elsewhere in relation to their use in planar chromatography (section 6.3.2). For now it suffices to say that capillary-controlled flow mechanisms are unsuitable for fast and efficient chromatographic separations. Pneumatic transport of the mobile phase is commonly employed in column chromatography. The mobile phase is pressurized externally and driven through the column by the pressure gradient between the column inlet and exit. A consequence of this pressure-driven flow is a parabolic radial velocity profile, Figure 1.5, and for compressible mobile phases, a local velocity that varies with position reflecting the decreasing flow resistance with migration along the column. Electroosmosis is the source of bulk liquid flow in an electric field. At the column wall or particle surface (packed columns) an electrical double layer results from the adsorption of ions from the

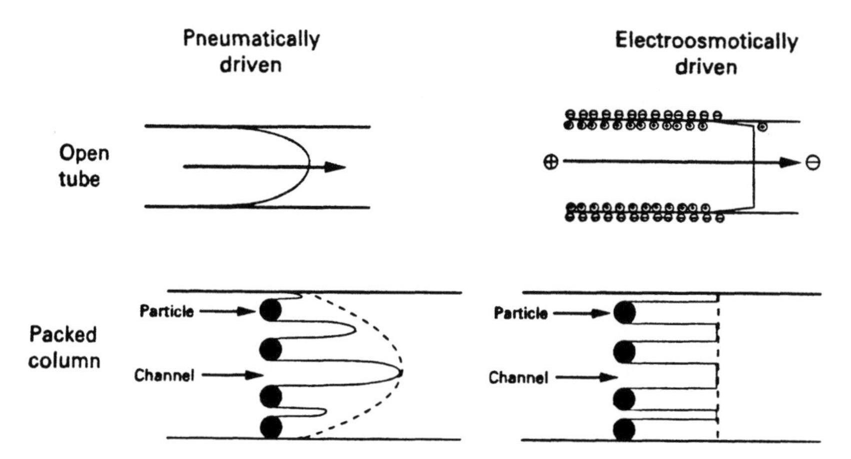


Figure 1.5. Mobile phase flow profile for an open tube and a packed column with pressure-driven and electroosmotic flow.

mobile phase or dissociation of surface functional groups. An excess of counterions is present in the double layer compared to the bulk liquid. In the presence of an electric field shearing of the solution occurs only within the microscopically thin diffuse part of the double layer transporting the mobile phase through the column with a nearly perfect plug profile, Figure 1.5. This is the general transport mechanism for neutral compounds in capillary electrophoresis, electrochromatography and micellar electrokinetic chromatography. The electrophoretic mobility of ions is also responsible for their migration in an electric field (section 8.2.1). An advantage of plug flow over parabolic flow is that it minimizes contributions from eddy diffusion to the column plate height, a major source of band broadening in pressure-driven chromatography. The electroosmotic velocity depends on a completely different set of column parameters to pressure-driven flow and for liquids provides an alternative transport mechanism to obtain efficient separation systems.

In pressure-driven flow Darcy's law provides the defining relationship between the column pressure drop and column characteristic properties. For gas chromatography, the mobile phase velocity at the column outlet is given by Eq. (1.12)

$$u_0 = KP_0(P^2 - 1) / 2\eta L \tag{1.12}$$

where K is the column permeability, $P_0$ the column outlet pressure, P the relative pressure (ratio of the column inlet to outlet pressure), $\eta$ the mobile phase viscosity and L the column length. Eq. (1.12) is valid for open tubular columns under all normal conditions and for packed columns at low mobile phase velocities. Since liquids are largely incompressible under normal operating conditions the equivalent relationship for liquid chromatography is Eq. (1.13) [114]

$$u = \Delta P K_0 d_p^2 / \eta L \tag{1.13}$$

where $\Delta P$ is the column pressure drop, $K_0$ the specific permeability coefficient, and $d_p$ the average particle diameter. Eq. (1.13) is valid for pressure drops up to about 600 atmospheres. The specific permeability coefficient has a value of about $1 \times 10^{-3}$, and can be estimated from the semi-empirical Kozeny-Carman equation [111]. The product $K_0 d_p^2$ is the column permeability. Eq. (1.12) and (1.13) are important for setting practical limits to the performance of pressure-driven chromatographic systems. Eventually inlet pressure becomes the upper bound for the achievable separation performance.

### 1.5.2 Rate Theories

Rather than a single defining relationship a number of similar models that encompass the same sources of band broadening but express the results in a different mathematical form are used in chromatography [12,37,103-108,115]. It was only quite recently that NMR imaging and photographic recording were used to visualize band broadening within the column packing [116-118]. Fortuitously these studies, and those performed using radial position detection, have served to demonstrate that basic theories are adequate to account for experimental observations. Interestingly, these studies have shown that dry packed and slurry packed columns exhibit different behavior with respect to radial velocity variations. Column permeability is highest in the core region for dry packed columns and at the wall for slurry packed columns. These studies confirm that velocity heterogeneity across the column radius combined with parabolic flow in pressure-driven chromatographic systems is a major cause of band broadening in packed columns.

Rate theory considers three general contributions to band broadening, which can be identified as eddy diffusion, longitudinal diffusion, and resistance to mass transfer. These contributions to the band broadening mechanism are treated as independent variables except under some circumstances when the eddy diffusion term is coupled to the mobile phase mass transfer term. The general approach can be applied to gas, liquid and supercritical fluid mobile phases, although it is necessary to make allowances for the differences in mobile phase physical properties, Table 1.5. The details of band broadening for supercritical fluid mobile phases are more complex than for gas and liquid mobile phases and to allow a simple presentation these are presented separately in section 7.4. Instrumental contributions to band broadening can be significant and are often unavoidable for miniaturized separation systems (section 1.5.4).

Eddy diffusion results from radial flow inequalities through a packed bed. The packing density for all real columns is heterogeneous in both the radial and axial direction. In addition, streamlines are not straight since molecules are forced to continually change direction by the obstacles (packing material) in their way. The local mobile phase velocity is faster through open spaces and wider interparticle spaces. The interparticle space is made up of a network of interconnected channels that experience very fast changes in cross section that depends on the particle shape and packing density. Because of parabolic flow in pressure-driven systems the mobile phase velocity is close to zero at the

Table 1.5
Characteristic values for column parameters related to band broadening

| Property | Column type | Mobile phase Gas | Supercritical fluid | Liquid |
|---|---|---|---|---|
| Diffusion coefficient ($m^2$ / s) | | $10^{-1}$ | $10^{-4}$ - $10^{-3}$ | $10^{-5}$ |
| Density (g / $cm^3$) | | $10^{-3}$ | 0.3 - 0.8 | 1 |
| Viscosity (poise) | | $10^{-4}$ | $10^{-4}$ - $10^{-3}$ | $10^{-2}$ |
| Column length (m) | Packed | 1 - 5 | 0.1 - 1 | 0.05 - 1 |
| | Open tubular | 10 - 100 | 5 - 25 | |
| Column internal diameter | Packed | 2 - 4 | 0.3 - 5 | 0.3 - 5 |
| (mm) | Open tubular | 0.1 - 0.7 | 0.02 - 0.1 | < 0.01 |
| Average particle diameter (μm) | | 100 - 200 | 3 - 20 | 3 - 10 |
| Column inlet pressure (atm) | | < 10 | < 600 | < 400 |
| Optimum velocity (cm / s) | Packed | 5 - 15 | 0.4 - 0.8 | 0.1 - 0.3 |
| | Open tubular | 10 - 100 | 0.1 - 0.5 | |
| Minimum plate height (mm) | Packed | 0.5 - 2 | 0.1 - 0.6 | 0.06 - 0.30 |
| | Open tubular | 0.03 - 0.8 | 0.01 - 0.05 | > 0.02 |
| Typical system efficiency (N) | Packed | $10^3$ - $10^4$ | 1 - 8 x $10^4$ | 0.5 - 5 x $10^4$ |
| | Open tubular | $10^4$ - $10^6$ | $10^4$ - $10^5$ | |

particle surface and increases rapidly towards the center of each channel. Thus, local velocities vary strongly, depending on the size of the interstices, the proximity to a particle surface, and the continuous blocking of the flow channels by successive particles. Molecular movement through such beds is possible only when molecules avoid the obstacles by moving round them in all possible radial and axial directions. However, because of radial diffusion individual solutes are not restricted to a single streamline but sample many in their passage through the packed bed. This results in an averaging of flow inequalities that acts to minimize dispersion. This is less than perfect and the uncompensated flow inequalities result in additional band broadening that would not exist if the packing structure was ideal. In the simplest sense these variations in the flow direction and rate lead to band broadening that should depend only on the density and homogeneity of the column packing. Its contribution to the total plate height, $H_E$, is proportional to the average particle size, $d_p$, and the column packing factor, $\lambda$, expressed by $H_E = 2\lambda d_p$. The packing factor is a dimensionless constant and usually has a value between 0.5 and 1.5. Band broadening due to eddy diffusion in packed beds can be minimized by employing packings of the smallest practical particle size with a narrow particle size distribution. The column pressure drop will ultimately determine the most practical particle size and column length. For typical operating pressures this corresponds to an average particle size of about 100 μm for gas chromatography and about 3 to 5 μm for liquid chromatography. For open tubular columns the eddy diffusion term is zero because there are no particles to disrupt flow lines.

The contribution to the plate height from molecular diffusion in the mobile phase arises from the natural tendency of the solute band to diffuse away from regions of high concentration to regions of lower concentration. Its contribution to the total plate height, $H_L$, is proportional to the diffusion coefficient in the mobile phase, $D_M$, the tortuosity (obstruction) factor of the column, $\gamma$, and the time the sample spends in the column according to $H_L = (2\gamma D_M / u)$. The obstruction factor is a dimensionless quasi-constant that is not totally independent of the mobile phase velocity. This dependence arises from the fact that the lowest flow resistance is offered by gaps or voids in the packing structure. Thus, at low velocities the value of the obstruction factor is averaged over tightly packed and loosely packed domains, while at high velocities it is weighted in favor of the loosely-packed domains where more flow occurs. Typical values for the obstruction factor are 0.6 to 0.8 in a packed bed and 1.0 for an open tubular column. Diffusion coefficients are about 10,000 times larger for gases than liquids and the contribution from longitudinal diffusion is almost always important in gas chromatography but is often negligible in liquid chromatography. Longitudinal diffusion is the principal source of band broadening in capillary electrophoresis and micellar electrokinetic chromatography (section 8.2.3).

Resistance to mass transfer is determined by the limitations of diffusion in the mobile and stationary phases as a transport mechanism to move analyte molecules to the boundary region between phases. Molecules in the vicinity of the phase boundary access the opposite phase quickly while those at a greater distance will require more time. During this time analyte molecules in the mobile phase will be transported some distance along the column. Since mass transfer is not instantaneous complete equilibrium is not established under normal separation conditions. The result is that the analyte concentration profile in the stationary phase is always displaced slightly behind the equilibrium position and the mobile phase profile is similarly slightly in advance of the equilibrium position. The combined peak observed at the column outlet is broadened about its band center, which is located where it would have been for instantaneous equilibrium, provided that the degree of non-equilibrium is small. The stationary phase contribution to resistance to mass transfer, $H_S$, is given by Eq. (1.14)

$$H_S = 2kd_f^2u / 3D_S(1 + k)^2 \qquad (1.14)$$

where $d_f$ is the stationary phase film thickness, and $D_S$ the diffusion coefficient in the stationary phase. Eq. (1.14) applies exactly to thin-film open tubular columns and is a reasonable approximation for packed column gas chromatography. For liquid chromatography the agreement is poor since there is no allowance made for the contribution of slow diffusion in the stagnant mobile phase. Here, Eq (1.15) provides a more realistic model for the stationary phase mass transfer contribution to the column plate height in liquid chromatography than Eq. (1.14)

$$H_S = [\theta(k_0 + k + k_0k)^2d_P^2u_e] / [30D_Mk_0(1 + k_0)^2(1 + k)^2] \qquad (1.15)$$

where $\theta$ is the tortuosity factor for the pore structure of the particles, $k_0$ the ratio of the intraparticle and interparticle volumes, and $u_e$ the interparticle mobile phase velocity. With multicomponent eluents the value of $k_0$ will vary with the mobile phase composition since the intraparticle space occupied by the stagnant mobile phase may change due to solvation of the stationary phase surface. In the derivation of Eq. (1.15), the influence of diffusion through the interparticle stagnant mobile phase has been neglected as it is generally very small compared to the value for the intraparticle stagnant mobile phase contribution.

The calculation of resistance to mass transfer in the mobile phase requires an exact knowledge of the flow profile of the mobile phase. This is only known exactly for open tubular columns for which the contribution to the total plate height from resistance to mass transfer in the mobile phase, $H_M$, can be described by Eq. (1.16)

$$H_M = [(1 + 6k + 11k^2) / 96(1 + k)^2][d_C{}^2u / D_M] \quad (1.16)$$

where $d_C$ is the column diameter. In a packed bed the mobile phase flows through a tortuous channel system and radial mass transfer can take place by a combination of diffusion and convection. The diffusion contribution to the total plate height from resistance to mass transfer in the mobile phase, $H_{M,D}$, is given approximately by Eq. (1.17)

$$H_{M,D} = wud_P{}^2 / D_M \quad (1.17)$$

where w is an empirical packing factor function used to correct for radial diffusion (ca. 0.02 to 5). To account for the influence of convection, that is, band broadening resulting from the exchange of solute between flow streams moving at different velocities, the eddy diffusion term must be coupled to the mobile phase resistance to mass transfer term, as indicated below

$$H_{MC} = 1 / (1/H_E + 1/H_{M,D}) \quad (1.18)$$

where $H_{MC}$ is the contribution to the plate height resulting from the coupling of eddy diffusion and mobile phase mass transfer terms. In general, $H_{MC}$ increases with increasing particle size and flow velocity and decreases with increasing solute diffusivity. The packing structure, the velocity range, and the retention factor can significantly influence the exact form of the relationship. In gas chromatography, the coupled plate height equation flattens out the ascending portion of the van Deemter curve at high mobile phase velocities in agreement with experimental observations. At flow velocities normally used the coupling concept appears to be unnecessary to account for experimental results. In liquid chromatography the existence of a coupling term and its most appropriate form is a matter that remains unsettled [37,107].

Although the above listing of contributions to the column plate height is not comprehensive, it encompasses the major band-broadening factors and the overall plate height can be expressed as their sum ($H_E + H_L + H_S + H_M$). A plot of the column plate height, H, against the mobile phase velocity is a hyperbolic function (Figure 1.6) most generally described by the van Deemter equation (1.19) [119].

$$H = A + B / u + (C_S + C_M)u \quad (1.19)$$

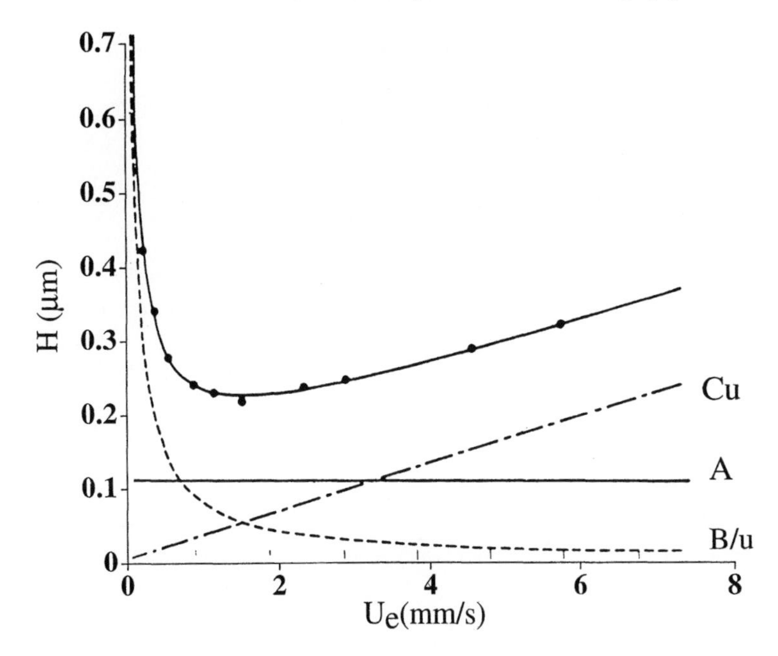


Figure 1.6. Relationship between the column plate height and mobile phase velocity for a packed column in liquid chromatography.

The A term represents the contribution from eddy diffusion, the B term the contribution from longitudinal diffusion, and the C terms the contributions from resistance to mass transfer in the stationary and mobile phases to the total column plate height. By differentiating equation (1.19) with respect to the mobile phase velocity and setting the result equal to zero, the optimum values of mobile phase velocity ($u_{opt}$) and plate height ($H_{min}$) can be obtained.

$$u_{opt} = [B / (C_M + C_S)]^{1/2} \tag{1.20}$$

$$H_{min} = A + 2[B(C_M + C_S)]^{1/2} \tag{1.21}$$

An important general contribution of the van Deemter equation was the illustration that an optimum mobile phase velocity existed for a column at which its highest efficiency would be realized. For less demanding separations columns may be operated at mobile phase velocities higher than $u_{opt}$ to obtain shorter separation times. Provided that the ascending portion of the van Deemter curve is fairly flat at higher velocities than $u_{opt}$, then the saving of time for a small loss in efficiency is justified. From Eq. (1.21) we see that the highest efficiency of a packed column is never less than the contribution from eddy diffusion so there is no redeeming value for columns with anything but a homogenous packing structure.

Optimization of column properties with respect to band broadening comes from an understanding and interpretation of the coefficients of the van Deemter equation [105,120-123]. For gas chromatography the compressibility of the mobile phase and its influence on the local velocity and diffusion coefficients along the column must be accounted for. Rearrangement in terms of the outlet pressure and velocity aids a general interpretation that is simple to do for the eddy diffusion term, longitudinal diffusion, and the mobile phase mass transfer term. The stationary phase mass transfer term, however, cannot be expressed in an explicit form independent of the column pressure drop. At high column pressure drops, such as those that might be encountered in fast gas chromatography with narrow bore open tubular columns, modification of Eq. (1.19) was recommended to include a change in the general dependence on the mobile phase gas velocity ($H = B / u^2 + C_M u^2 + C_S u$) [124].

Beginning with the most favorable case, band broadening in open tubular columns is satisfactorily described by the Golay equation, extended to situations of appreciable pressure drop by Giddings, Eq. (1.22)

$$H = f_1[(2D_{M,o} / u_o) + (f_g(k))(d_C{}^2 u_o / D_{M,o})] + f_2[(f_S(k))d_f{}^2 u_o / D_S] \qquad (1.22)$$

$$f_1 = 9/8\,(P^4 - 1)(P^2 - 1) / (P^3 - 1)^2$$

$$f_2 = 3/2\,(P^2 - 1) / (P^3 - 1)$$

$$f_g(k) = (1 + 6k + 11k^2) / 96\,(1 + k)^2$$

$$f_S(k) = 2k / 3(1 + k)^2$$

where $D_{M,o}$ is the mobile phase diffusion coefficient at the column outlet pressure, $u_o$ the mobile phase velocity at the column outlet, $d_f$ the stationary phase film thickness, and P the ratio of column inlet to outlet pressure [122-126].

Open tubular columns in current use have internal diameters within the range 0.1 to 0.6 mm and a stationary phase film thickness from about 0.05 to 8.0 $\mu$m. Gases of high diffusivity, hydrogen or helium, are used to minimize mass transfer resistance in the mobile phase and, at the same time, minimize separation time ($H_{min}$ occurs at higher values of $u_{opt}$ for gases of high diffusivity). Narrow bore columns are capable of higher intrinsic efficiency since they minimize the contribution from resistance to mass transfer in the mobile phase to the column plate height, Figure 1.7. The $C_M$ term increases successively with increasing column diameter and is also influenced by the retention factor, particularly at low values of the retention factor. When combined with the term describing the plate height contribution due to longitudinal diffusion, $C_M$ is the dominant cause of band broadening for wide bore, thin-film columns. The stationary phase mass transfer term becomes increasingly important as film thickness increases, Table 1.6 [127]. For thin-film columns ($d_f < 0.25$ $\mu$m) the stationary phase resistance to mass transfer term is generally only a few percent of the mobile phase

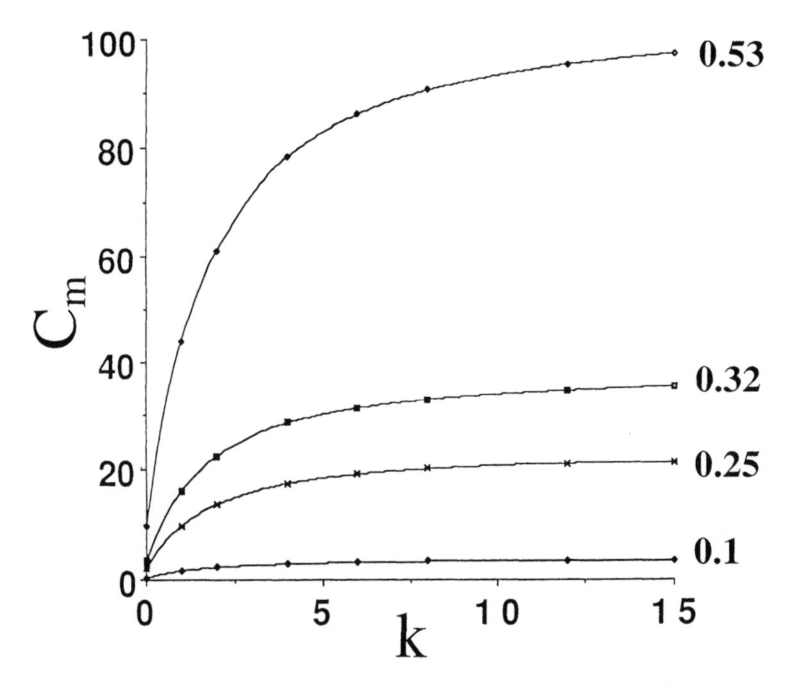


Figure 1.7. Variation of the resistance to mass transfer in the mobile phase, $C_M$, as a function of the retention factor for open tubular columns of different internal diameters (mm).

Table 1.6
Relative contribution (%) of resistance to mass transfer in the mobile and stationary phases to the column plate height for undecane at 130°C for a 0.32 mm internal diameter open tubular columns in gas chromatography

| Film | Retention | Phase | Mass transfer term (%) | |
|---|---|---|---|---|
| thickness (μm) | factor | ratio | $C_M$ | $C_S$ |
| 0.25 | 0.56 | 320 | 95.2 | 4.8 |
| 0.5 | 1.12 | 160 | 87.2 | 12.8 |
| 1.00 | 2.24 | 80 | 73.4 | 26.6 |
| 5.00 | 11.2 | 16 | 31.5 | 68.5 |

term and, to a first approximation, can be neglected. In estimating the contribution of stationary phase mass transfer resistance to the plate height there is a strong dependence on the retention factor and the diffusion coefficient of the solute in the stationary phase, Figure 1.8. Diffusion coefficients in polar, gum and immobilized phases tend to be smaller than those observed for other phases. Thick-film open tubular columns prepared from polar immobilized stationary phases tend to be significantly less efficient than similar columns prepared from low polarity stationary phases; the efficiency of both column types decreases with increasing film thickness.

The Golay equation can also be used to predict optimum separation conditions in open tubular column liquid chromatography [128,129]. The main difference between

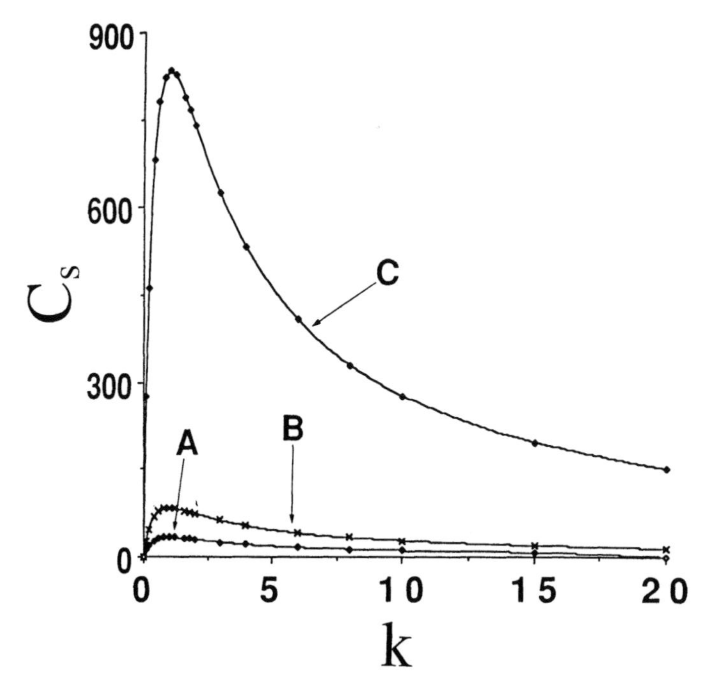


Figure 1.8. Variation of the resistance to mass transfer in the stationary phase, $C_S$, as a function of the retention factor for different film thicknesses. A, $d_f$ = 1 μm and $D_S$ = 5 x $10^{-7}$ $cm^2/s$; B, $d_f$ = 5 μm and $D_S$ = 5 x $10^{-6}$ $cm^2/s$; and C, $d_f$ = 5 μm and $D_S$ = 5 x $10^{-7}$ $cm^2/s$.

gas and liquid chromatography in open tubular column is that the diffusion coefficients in liquids are roughly 10,000 times smaller than in gases and therefore the last term in Eq. (1.22) can be neglected. For high efficiency the column internal diameter must be reduced to a very small size (1-10 μm) to overcome the slow mass transfer in the mobile phase. This creates considerable instrument and column technology constraints that limit the practical utility of open tubular column liquid chromatography at present.

Since the exact flow profile of the mobile phase through a packed bed is unknown, only an approximate description of the band broadening process can be attained. For packed column gas-liquid chromatography at low mobile phase velocities, Eq. (1.23) provides a reasonable description of the band broadening process [103].

$$H = 2\lambda d_P + 2\gamma D_{M,o} / u_o + [f_g(k)](d_P^2 / D_{M,o})u_o + [f_S(k)](d_f^2 / D_S)u \qquad (1.23)$$

According to Scott the average linear velocity can be replaced by $(4u_o / [P + 1])$ in Eq. (1.23) to permit evaluation entirely in terms of the outlet velocity [130]. If $\lambda = 0$, $\gamma = 1$, and $d_P = r_c$ is substituted into Eq. (1.23) then this equation can be used as an alternative to Eq. (1.22) to account for band broadening in evaluating open tubular columns [121,130]. For gas-solid chromatography the stationary phase mass transfer term ($C_S$) is replaced by $C_k$ describing the kinetics of adsorption and desorption from a solid surface, which is often the dominant term in the plate height equation for inorganic oxide and chemically bonded adsorbents [131,132].

For packed column gas chromatography small particles with a narrow size distribution and coated with a thin, homogeneous film of liquid phase are required for high efficiency. The typical range of inlet pressures controls the absolute particle size. Column packings with particle diameters of 120-180 μm in columns less than 5 meters long are generally used with inlet pressures less than 10 atmospheres. For a liquid phase loading of 25-35% w/w, slow diffusion in the stationary phase film is the principal cause of band broadening. With lightly loaded columns (< 5% w/w), resistance to mass transfer in the mobile phase is no longer negligible. At high mobile phase velocities the coupled form of the plate height equation is used to describe band broadening.

Several equations in addition to the van Deemter equation, have been used to describe band broadening in liquid chromatography, Eqs. (1.23) to (1.26) [104,106,107,132-134].

$$H = A / [1 + (E / u)] + B / u + Cu \tag{1.23}$$

$$H = A / [1 + (E / u^{1/2})] + B / u + Cu + Du^{1/2} \tag{1.24}$$

$$H = Au^{1/3} + B / u + Cu \tag{1.25}$$

$$H = A / [(1 + E / u^{1/3})] + B / u + Cu + Du^{2/3} \tag{1.26}$$

A, B, C, D, and E are appropriate constants for a given solute in a given chromatographic system. A comparison of these equations by Katz et al. [135] indicated a good fit with experimental data for all equations, but only Eqs. (1.19), (1.23), and (1.25) consistently gave physically meaningful values for the coefficients.

The highest efficiency in liquid chromatography is obtained using columns packed with particles of small diameter, operated at high inlet pressures, with mobile phases of low viscosity. Diffusion coefficients are much smaller in liquids than in gases and, although this means that longitudinal diffusion as a source of band broadening can often be neglected, the importance of mass transfer in the mobile phase is now of greater significance. The adverse effect of slow solute diffusion in liquid chromatography can be partially overcome by operating at much lower mobile phase velocities than is common for gas chromatography. This increase in efficiency, however, is obtained at the expense of longer separation times. Slow mass transfer in the stationary phase is a result of slow diffusion through the stagnant mobile phase trapped in the pore structure and surface diffusion along the particle surface of solvated chemically bonded phases [136]. Unfavorable intraparticle mass transfer in porous particles is minimized by using particles of smaller diameter, Figure 1.9, since this restricts the average path length over which the solute must be transported by diffusion. For particles less than 5 μm in diameter the plate height curves are essentially flat in the region of the minimum value indicating

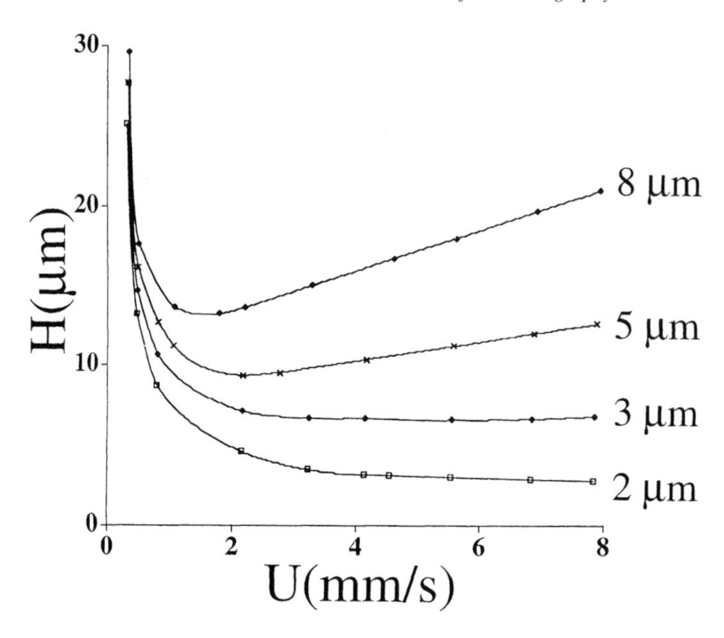


Figure 1.9. Plot of the plate height H (μm) against the mobile phase velocity u (mm/s) for columns of different particle diameters (and different column lengths) in liquid chromatography.

that small-particle diameter columns can be operated at higher linear velocities without appreciable loss in efficiency. The high flow resistance of these columns, however, prevents long columns from being used with typical inlet pressures. Nonporous particles, of course, provide an optimum means of minimizing the contribution of slow intraparticle mass transfer to column efficiency in liquid chromatography [137]. The low surface areas of these particles, however, are a disadvantage for optimizing the separation of small molecules owing to their low retention. For macromolecules this is less of a problem. The diffusion coefficients of macromolecules are one to two orders of magnitude smaller than for low molecular weight compounds. For macromolecules small diameter porous and nonporous particles yield enhanced column efficiency by virtue of the relatively small intraparticle mass transfer term (short diffusion distances) and to a lesser extent due to the small contribution of eddy diffusion to the plate height [138]. High temperatures [139] and enhanced fluidity mobile phases (prepared by adding low viscosity liquids, usually carbon dioxide, to typical mobile phases) [140,141] improve solute diffusion properties and decrease mobile phase viscosity. This results in increased efficiency and shorter separation times.

### 1.5.3 Reduced Parameters

The major advantage of using reduced parameters (h, $\upsilon$, $\phi$) in place of the absolute parameters (H, u, $K_0$) is that results obtained from columns containing packing materials of different sizes, operated with mobile phases of different viscosity, and evaluated with solutes of different diffusion coefficients, can be compared directly. This approach also leads to a simple index of performance, the separation impedance, which can be used to judge the relative performance of columns of a similar kind as well as the limit to performance of different column types. Reduced parameters are widely used in all types of chromatography but it was in the early development of liquid chromatography that they had their main impact on column design and testing [107,108,142-146]. Since liquids are largely incompressible the use of reduced parameters in liquid chromatography is straightforward. For compressible mobile phases it is necessary to account for the influence of the operating conditions on mobile phase viscosity and diffusion coefficients. For simplicity the examples employed in this section will be for liquid chromatography but the general approach is sound for any chromatographic technique except for those based on electrophoretic migration.

The reduced plate height, h, is defined as the number of particles to a theoretical plate and is given by

$$h = H / d_P = (1 / 5.54)(L / d_P)(w_h / t_R)^2 \quad (1.27)$$

where $w_h$ is the peak width at half height and $t_R$ the solute retention time. The reduced velocity, $\upsilon$, is the rate of flow of the mobile phase relative to the rate of diffusion of the solute over one particle diameter and is given by

$$\upsilon = u d_P / D_M = L d_P / t_M D_M \quad (1.28)$$

where $D_M$ is the solute diffusion coefficient in the mobile phase and $t_M$ the column holdup time. When the diffusion coefficient is unknown an approximate value can be found from the Wilke-Chang equation

$$D_M = A(\psi M_2)^{1/2} T / \eta V^{0.6} \quad (1.29)$$

where A is a constant depending on the units used (equal to 7.4 x $10^{-12}$ when $D_M$ is in $m^2/s$), $\psi$ a solvent dependent constant (1.0 for unassociated solvents, 1.5 for ethanol, 1.9 for methanol, and 2.6 for water), T the temperature in K, $\eta$ the mobile phase viscosity, and V the solute molar volume [147]. For mixed solvents the volume average value for the product ($\psi M_2$) is used [145]. Typical values for low molecular weight solutes fall into the range 0.5-3.5 x $10^{-9}$ $m^2/s$. The higher value is typical of organic solvents of low viscosity such as hexane, and the lower value for polar aqueous solvents. The column flow resistance parameter, $\phi$, is a measure of the resistance to flow of the mobile phase and takes into account the influence of the column length, particle diameter, and mobile phase viscosity. It is given by

$$\phi = \Delta P d_P^2 t_M / \eta L^2 \quad (1.30)$$

Table 1.7
Typical values of the reduced parameters for liquid chromatography columns

| Column type | Minimum reduced plate height | Minimum reduced velocity | Flow resistance parameter | Separation impedance |
|---|---|---|---|---|
| Conventional packed | 1.5-3.0 | 3-5 | 500-1000 | 2000-9000 |
| Small-bore packed | 1.5-3.0 | 3-5 | 500-1000 | 2000-9000 |
| Packed capillary | 2.0-3.5 | 1-5 | 350-1000 | 3000-7000 |
| Open Tubular | 0.5-30 | 4-30 | 32 | 8-80 |

where $\Delta P$ is the column pressure drop. Finally, the separation impedance, E, which represents the elution time per plate for an unretained solute times the pressure drop per plate, the whole corrected for viscosity is given by

$$E = (t_M / N)(\Delta P / N)(1 / \eta) = h^2\phi \quad (1.31)$$

The separation impedance represents the difficulty of achieving a certain performance and should be minimized for optimum performance. The highest performance is achieved by a column, which combines low flow resistance and produces minimal band broadening. For open tubular columns the column internal diameter, $d_C$, replaces $d_P$ in Eqs. (1.27), (1.28) and (1.30).

Some typical reduced parameter values for different liquid chromatography columns are summarized in Table 1.7. A reduced plate height of 2 is considered excellent for a packed column with more typical values for good columns lying between 2 and 3. The flow resistance parameter for packed columns will normally lie within the range 500 to 1000. It provides information on how the chromatographic system as a whole is performing. Unusual values of the flow resistance parameter are often associated with blockages in the system (frits, connecting tubing, etc.), packings containing fine particles, or columns containing excessive voids. The separation impedance then has values between 2000 and 9000 for good columns. The performance of packed columns should be independent of the column diameter and this is reflected in the similar values for the separation impedance for conventional, small bore, and packed capillary columns. Slightly lower plate heights and reduced flow resistance have been observed for packed capillary columns with low aspect ratios (ratio of column diameter / particle diameter) [148-150]. This is most likely due to the influence of the column wall on the packing density resulting in decreased flow dispersion and a reduction in the mobile phase mass transfer contribution to the plate height.

According to chromatographic theory, the reduced plate height is related to the reduced velocity by Eq. (1.32)

$$h = A\upsilon^{1/3} + B / \upsilon + C\upsilon \quad (1.32)$$

The constant B reflects the geometry of the mobile phase in the column and the extent to which diffusion of the solute is hindered by the presence of the packing. For columns yielding acceptable fits to Eq. (1.32), B is expected to lie between 1 and 4, and for small molecules is typically around 2. It is responsible for the decrease in efficiency at very low flow rates. The constant A is a measure of the uniformity and density of the column packing. A well-packed column will have a value of A between 0.5 and 1.0 while a poorly packed column will have a higher value, say between 2 and 5. The constant C reflects the efficiency of mass transfer between the stationary phase and the mobile phase. At high reduced mobile phase velocities the C term dominates the reduced plate height value, and therefore column efficiency. The value for C is close to zero for a nonporous sorbent, a reasonable value is 0.003, but has a greater value for silica-based porous sorbents with a value of 0.05 being reasonable for the latter. Porous polymers may have significantly higher values approaching unity.

The constant A, B, and C can be determined by curve fitting from a plot of the reduced plate height against the reduced velocity [145,151-153]. Accurate values for the constants are only obtained if a wide range of reduced velocity values is covered and the data are of high quality. From the shape of a log-log plot of reduced plate height against reduced velocity, Figure 1.10, the important features of the column are easily deduced. If the minimum is below 3 and occurs in the range $3 < \upsilon < 10$ then the column is well packed (low A). If the reduced plate height is below 10 at a reduced velocity of 100 then the material has good mass transfer characteristics (low C) and is probably well packed as well (low A). If the curve has a high and flat minimum the column is poorly packed (high A).

Because of the time required to develop sufficient data points to make a plot similar to that shown in Figure 1.10 it is useful to have a shorter method for assessing potential problems. For a good column the value of the reduced plate height should not exceed 3 or 4 at a reduced velocity of about 5 and 10 to 20 at a reduced velocity of about 100.

The possibility of obtaining significant improvements in performance by using open tubular columns in liquid chromatography is clearly illustrated by the values for the separation impedance in Table 1.7. The dependence of the reduced plate height of an open tubular column on the reduced velocity is expressed by Eq. (1.33), assuming that the resistance to mass transfer in the stationary phase can be neglected

$$h = H / d_C = (2 / \upsilon) + [(1 + 6k + 11k^2) / 96(1 + k)^2]\upsilon \qquad (1.33)$$

The reduced plate height is reasonably constant, independent of the retention factor for a packed column, while Eq. (1.33) indicates that the reduced plate height, at least for small values of the retention factor, will increase with the retention factor for open tubular columns. Actual values of the minimum reduced plate height will vary from about 0.3 (corresponding to $k = 0$) to 1.0 (corresponding to $k = \infty$) for values of the reduced velocity in the range 4-14 [142]. The exceptional potential performance of open tubular columns cannot be explained entirely by the smaller values of the reduced plate height. The most significant difference compared to a packed column is their greater permeability. The column flow resistance parameter can be calculated directly from the Poisseuille equation and is exactly 32 for open tubular columns [143]. The separation

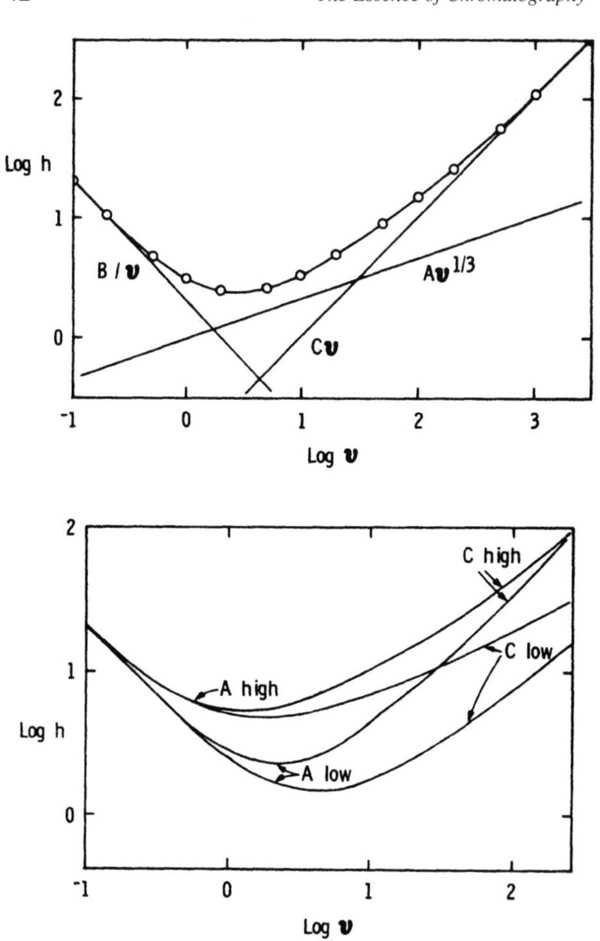


Figure 1.10. Plot of the reduced plate height against the reduced mobile phase velocity indicating the velocity region in which the three terms in Eq. (1.31) contribute to the reduced plate height. The lower figure shows some typical results from column testing. The lowest curve (A and C low) illustrates ideal behavior. The plots for other columns can be considered to deviate from this curve due to high A (poorly packed column), high C (poor quality column packing), or from having both a high A and C value.

impedance is about two orders of magnitude lower for an open tabular column compared to a packed column. If operation at the minimum separation impedance was possible then for a separation requiring a fixed number of plates the open tubular column would be about 100 times faster at a constant inlet pressure

The optimum operating conditions for the various column types are summarized in Table 1.8 [142]. There are no serious problems when operating conventional packed columns under optimum conditions, at least for particles > 2 μm. Small bore packed columns place greater demands on instrumentation but again it should be feasible to

Table 1.8
Optimum operating conditions for different columns in liquid chromatography
$\Delta P = 200$ bar, $\eta = 1 \times 10^{-3}$ N.s/m$^2$ and $D_M = 1.0 \times 10^{-9}$ m$^2$/s

| Column plate count | Hold-up time (s) | Particle or column diameter (μm) | Length (m) | Flow rate (μl/s) | Peak standard deviation k = 0 (μl) |
|---|---|---|---|---|---|
| Conventional Packed Columns: $d_C = 5$ mm, $\phi = 500$; $h_{min} = 2$; $\upsilon_{opt} = 5$; $E_{min} = 2000$; $\varepsilon_T = 0.75$ | | | | | |
| 10,000 | 10 | 1.6 | 0.03 | 50 | 5 |
| 30,000 | 90 | 2.7 | 0.17 | 30 | 15 |
| 100,000 | 1,000 | 5 | 1 | 15 | 50 |
| 300,000 | 9,000 | 9 | 5 | 9 | 150 |
| 1,000,000 | 100,00 | 16 | 3 | 5 | 500 |
| Small Bore Packed Columns: $d_C = 1$ mm, $\phi = 500$; $h_{min} = 2$; $\upsilon_{opt} = 5$; $E_{min} = 2000$; $\varepsilon_T = 0.75$ | | | | | |
| 10,000 | 10 | 1.6 | 0.03 | 1.9 | 0.2 |
| 30,000 | 90 | 2.7 | 0.17 | 1.1 | 0.6 |
| 100,000 | 1,000 | 5 | 1 | 0.6 | 2 |
| 300,000 | 9,000 | 9 | 5 | 0.35 | 6 |
| 1,000,000 | 100,000 | 16 | 30 | 0.2 | 20 |
| Open Tubular Columns: $\phi = 32$; $h_{min} = 0.8$; $\upsilon_{opt} = 5$; $E_{min} = 20$ | | | | | |
| 10,000 | 0.1 | 0.25 | 0.002 | $1.0 \times 10^{-6}$ | $1.0 \times 10^{-9}$ |
| 30,000 | 0.9 | 0.43 | 0.01 | $1.6 \times 10^{-6}$ | $0.8 \times 10^{-8}$ |
| 100,000 | 10 | 0.8 | 0.065 | $3.0 \times 10^{-6}$ | $1.0 \times 10^{-7}$ |
| 300,000 | 90 | 1.4 | 0.33 | $5.0 \times 10^{-6}$ | $0.8 \times 10^{-6}$ |
| 1,000,000 | 1,000 | 2.5 | 2 | $1.0 \times 10^{-5}$ | $1.0 \times 10^{-5}$ |
| 3,000,000 | 9,000 | 4.3 | 10 | $1.6 \times 10^{-5}$ | $8.0 \times 10^{-5}$ |
| 10,000,000 | 100,000 | 8 | 63 | $3.0 \times 10^{-5}$ | $1.0 \times 10^{-3}$ |

meet the requirements for operation under optimum conditions for plate counts in excess of 10,000. There is virtually no possibility of operating open tubular columns under optimum conditions since the eluted peak volumes for the unretained solute are so extremely small, even for high plate counts. Extracolumn contributions to dispersion should be about one-third of the volume standard deviation of an unretained solute to avoid a significant loss in performance for early eluting peaks. For an open tubular column exhibiting more than 1 million plates this corresponds to an extracolumn volume of less than 10 picoliters. Also, in practice, it may be difficult to achieve the optimum column internal diameter which for a column exhibiting 1 million plates is about 2.5 μm. The data in Table 1.8 were compiled at a fixed pressure of 200 bar. Increasing the inlet pressure reduces the separation time but leads to smaller optimum dimensions for the column diameter and length. This in turn leads to a smaller value for the peak standard deviation and a more hopeless case in terms of the practicality of operating open tubular columns in liquid chromatography at the minimum separation impedance.

Since packed columns can be operated at $E_{min} = 2000$ with plate numbers from 10,000 to 1 million, alternative column types will compete effectively only if they can be operated at separation impedance values below 2000. Enlarging the column

Table 1.9
Detector limited operation of open tubular columns in liquid chromatography
$\eta = 10^{-3}$ N.s/m$^2$; $D_M = 10^{-9}$ m$^2$/s; $\Delta P$ = 200 bar; $\phi$ = 32; mass transfer coefficient = 0.08

| Column plate count | Column hold-up time (s) | Column length (m) | Reduced plate height | Reduced velocity | Separation impedance |
|---|---|---|---|---|---|
| $d_C$ = 27 μm and peak standard deviation (k = 0) = 0.1 μl | | | | | |
| 10,000 | 600 | 17 | 64 | 797 | 130,000 |
| 30,000 | 1,800 | 30 | 37 | 464 | 44,000 |
| 100,000 | 6,000 | 55 | 20 | 252 | 13,000 |
| 300,000 | 18,000 | 90 | 11 | 140 | 4,000 |
| 1,000,000 | 60,000 | 170 | 6.4 | 80 | 1,300 |
| $d_C$ = 9 μm and peak standard deviation (k = 0) = 1 nl | | | | | |
| 10,000 | 60 | 1.7 | 18.5 | 232 | 11,000 |
| 30,000 | 180 | 3 | 11.2 | 140 | 4,000 |
| 100,000 | 600 | 5.5 | 6.1 | 77 | 1,200 |
| 300,000 | 1,800 | 9 | 3.4 | 42 | 360 |
| 1,000,000 | 6,000 | 17 | | | 120 |

diameter and using higher reduced velocities will increase the separation impedance for the column, but still below 2000, while simultaneously increasing peak volumes to relax the instrumental constraints for column operation. Optimization under conditions where extracolumn volumes are limiting can be handled by setting a minimum value for the peak standard deviation of an unretained solute. A total instrument dispersion due to extracolumn volumes of about 0.1 μl could probably be achieved whereas 1 nl or smaller would certainly be more desirable. Considering the above considerations the separation impedance can be calculated for various sets of practical operating conditions as shown in Table 1.9. For extracolumn volume peak standard deviations of 0.1 μl an open tubular column will only show superiority over a packed column operated under optimum conditions when the plate count is in excess of about 500,000. With a further reduction to 1 nl the open tubular column is superior to the packed column for a plate count in excess of about 70,000. Consequently, the future of open tubular column liquid chromatography will depend primarily on the development of new instrument concepts with injector and detector volumes reduced to the nl level or less. Also, new column technology that permits the fabrication of narrow bore columns with sufficient stationary phase to provide reasonable partition ratios will be required. This is not a trivial accomplishment and it would seem unlikely that open tubular columns will be widely used in analytical laboratories at any time in the near future [154,155].

### 1.5.4 Extracolumn Sources

Under ideal conditions, the peak profile recorded during a separation should depend only on the operating characteristics of the column and should be independent of the instrument in which the column resides. Under less than ideal conditions, the peak profile will be broader than the column profile by an amount equivalent to the extracolumn band broadening. This broadening results from the volumetric dispersion originating from the injector, column connecting tubing and detector together with the temporal dispersion resulting from the slow response of the electronic circuitry of the detector and data recording device. The various contributions to extracolumn band broadening can be treated as independent factors, additive in their variances, according to Eq. (1.34) [156-161]

$$\sigma^2_T = \sigma^2_{col} + \sigma^2_{inj} + \sigma^2_{con} + \sigma^2_{det} + \sigma^2_{tc} \qquad (1.34)$$

where $\sigma^2_T$ is the total peak variance observed in the chromatogram, $\sigma^2_{col}$ the peak variance due to the column, $\sigma^2_{inj}$ the variance due to the volume and geometry of the injector, $\sigma^2_{con}$ the variance due to connecting tubes, unions, etc., $\sigma^2_{det}$ the variance due to the volume and geometry of the detector, and $\sigma^2_{tc}$ the variance due to the finite response time of the electronic circuits of the detector and data system. For experimental evaluation Eq. (1.34) can be simplified to $\sigma^2_T = \sigma^2_{col} + \sigma^2_{ext}$ where $\sigma^2_{ext}$ is the sum of all extracolumn contributions to the peak variance and represents the instrumental contribution to band broadening. The column contribution to the peak variance can be written in time or volume units, $\sigma^2_{col} = (t_R^2 / N) = (V_R^2 / N)$ where $t_R$ and $V_R$ are the retention time and volume, respectively, for a peak, and N the true column efficiency in the absence of extracolumn band broadening. A commonly accepted criterion for the instrumental contribution to band broadening is that this should not exceed 10% of the column variance. This corresponds to a loss of 10% of the column efficiency and about 5% of the column resolution. A peak eluting from a column will occupy a volume equivalent to 4 $\sigma$ units. The above criterion can then be used to establish working limits for the acceptable extracolumn variance and volume for typical separation conditions using the properties of an unretained peak as the worst case or limiting condition.

Band broadening due to injection arises because the sample is introduced into the column as a finite volume over a finite time. A solute zone is formed at the column head, which reflects the degree of sample axial displacement during the injection time. This solute zone is generally less than the injection volume due to retention of the solute by the stationary phase. In liquid and supercritical fluid chromatography the elution strength of the injection solvent and the effect of solvent dilution with the mobile phase will determine the extent of the zone displacement. This situation is too complex to be described by a simple mathematical model. Two extreme views, those of plug or exponential injection, can be used to define the limiting cases. Valve injection occurs largely by displacement (plug injection) accompanied by various contributions from exponential dilution (mixing). The variance due to the injection profile is described by $\sigma^2_{inj} = V^2_{inj} / K$ where $V_{inj}$ is the injection volume and K is a constant with values between 1-12 depending on the characteristics of the injector [156,159,162,163]. For plug injection K = 12; more typical values for K under conditions of valve injection in liquid chromatography are 2-8.

Band dispersion in open tubes is due to poor radial mass transfer of the solute resulting from the parabolic velocity profile that exists in cylindrical tubes. The solute contained in the mobile phase close to the wall is moving very slowly and that at the center at the maximum velocity. This range of solvent velocity from the wall of the tube to the center causes a significant increased dispersion of any solute band passing through it. The variance due to connecting tubes is given by

$$\sigma^2_{con} = \pi r_c^4 L / 24 D_M F \quad (1.35)$$

where $r_c$ is the tube radius, L the tube length, $D_M$ the solute diffusion coefficient in the mobile phase and F the column flow rate [106,157,161]. The dispersion in connecting tubes can be minimized by using short lengths of tubing with small internal diameters. Tubing with too small a diameter, however, increases the risk of plugging, and tubing of too short a length limits the flexibility for spatial arrangement of the instrument modules. Serpentine tubing, which introduces radial convection to break up the parabolic flow by reversing the direction of flow at each serpentine bend, is a practical alternative [164,165]. Dispersion in a serpentine tube is about 20% of that for a straight tube and allows longer connecting tubes to be used. Strictly speaking Eq. (1.35) requires a certain minimum efficiency to be an accurate description of the band broadening process. This minimum efficiency may not be reached in all cases, and the use of Eq. (1.35) may over estimate the connecting tube contribution to band broadening [166].

Depending on the design of the detector cell, it can have either the properties of a tube with plug flow or act as a mixing volume. In practice, most detector cells behave in a manner somewhere in between these two extreme models. When plug flow is dominant the variance can be calculated from $\sigma^2_{det} = V^2_{det} / 12$ and when mixing dominates from Eq. (1.35) by substituting the appropriate terms for the detector cell radius and length. Generally, if the cell volume is approximately 10% of the peak volume, then extracolumn band broadening from the detector will be insignificant. Higher flow rates through the detector cell result in a decrease in the peak residence time accompanied by a smaller contribution from extracolumn band broadening at the expense of sensitivity. Detectors and data systems can also cause band broadening due to their response time, which is primarily a function of the time constant associated with the filter network used to diminish high frequency noise. With the exception of fast chromatography the contribution of the electronic response time to band broadening should not be significant for modern instruments.

There are two general experimental methods for estimating the extracolumn band broadening of a chromatographic instrument. The linear extrapolation method is relatively straightforward to perform and interpret but rests on the validity of Eq. (1.34) and the model used to calculate the contribution for the column variance. A plot of $\sigma^2_T$ against $t_R^2$, $V_R^2$ or $(1 + k)^2$ for a series of homologous compounds will be linear. The true column efficiency can be obtained from the slope of the line and $\sigma^2_{ext}$ from the intercept on the vertical axis [162,167,168]. The assumption that the individual contributions to the extracolumn variance are independent may not be true in practice, and it may be necessary to couple some of the individual contributions to obtain the most accurate values for the extracolumn variance [156]. The calculation of the column contribution assumes that solute diffusion coefficients in the mobile and stationary phases are identical and that any variation in the column plate height, as a function of the retention factor, can be neglected [168]. These assumptions are unlikely to be true in all cases diminishing the absolute accuracy with which extracolumn variance can be calculated using the linear extrapolation method. The zero length column method uses a short length of connecting tubing to replace the column [106,157]. It is then assumed that the observed total dispersion for an injected peak arises from system components only. This approach is experimentally demanding and ignores the contribution from column connecting fittings to the extracolumn variance.

### 1.5.5 Isotherm Effects

For separation conditions pertaining to analytical chromatography it is often assumed that the distribution constant is independent of the analyte concentration and a plot of the analyte concentration in the stationary phase against its concentration in the mobile phase is linear with a slope equal to the distribution constant. Such a plot is called an isotherm because it is obtained at a single temperature. The resultant peak shape is symmetrical with a width dependent on the kinetic properties of the column. This is the basis of linear chromatography on which the general theory of chromatography is constructed. Nonlinear isotherms are observed under certain circumstances and result in peak asymmetry and retention times that depend on the concentration of analyte in the mobile phase, and therefore the injected sample sizes [27,169]. Nonlinear isotherms are common in preparative-scale chromatography due to the use of large sample sizes to maximize yield and production rate [12].

Common causes of nonlinear isotherms are high sample concentrations and energetically heterogeneous adsorbents containing adsorption sites with incompatible association/dissociation rate constants [12,21,169-171]. Chemically bonded phases used in liquid chromatography, for example, contain sorption sites associated with the surface bonded ligands and accessible silanol groups of the silica matrix [171-174]. The interaction of these sites with hydrogen-bonding solutes is likely to be different. Inorganic oxide sorbents used in gas-solid chromatography are also intrinsically heterogeneous [170]. For differences that are not too extreme experimental isotherms for monolayer coverage of solutes with strong solute-stationary phase interactions and weak solute-solute interactions can be fit to a Langmuir-type model, Eq. (1.36)

$$Q = aC / (1 + bC) \qquad (1.36)$$

where Q is the analyte concentration in the stationary phase, C the analyte concentration in the mobile phase and a and b are experimental constants. The a constant is related to the distribution constant for the analyte at infinite dilution and the b constant the sorbent

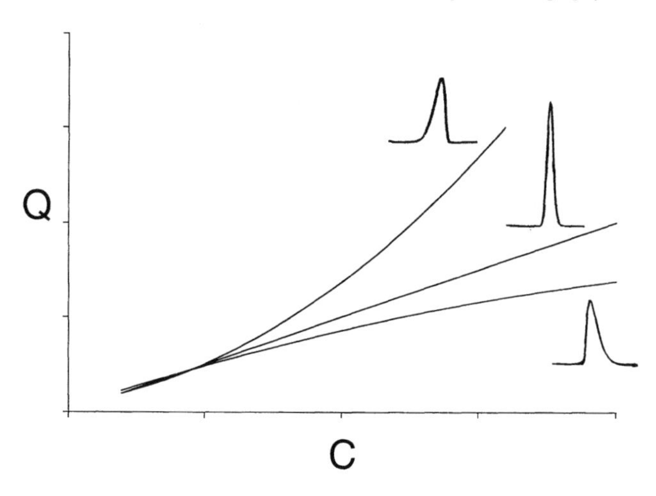


Figure 1.11. Schematic representation of different isotherms and their influence on chromatographic peak shape. Q is the analyte concentration in the stationary phase and C the analyte concentration in the mobile phase.

saturation capacity. For isotherms that do not fit Eq. (1.36) alternatives include the bi-Langmuir, Freundlich, or quadratic models [12,170,176]. In the case of a Langmuir-type isotherm, the number of unoccupied sorption sites on the stationary phase are rapidly reduced as the analyte concentration increases and the slope of the isotherm decreases, Figure 1.11. As a result, at higher analyte concentrations the band moves faster than at lower concentrations and the peak develops a tail at its rear.

Anti-Langmuir type isotherms are more common in partition systems where solute-stationary phase interactions are relatively weak compared with solute-solute interactions or where column overload occurs as a result of large sample sizes. In this case, analyte molecules already sorbed to the stationary phase facilitate sorption of additional analyte. Thus, at increasing analyte concentration the distribution constant for the sorption of the analyte by the stationary phase increases due to increased sorption of analyte molecules by those analyte molecules already sorbed by the stationary phase. The resulting peak has a diffuse front and a sharp tail, and is described as a fronting peak.

A number of experimental techniques have been described for the determination of isotherms based on frontal analysis, frontal analysis by characteristic point, elution by characteristic point, and perturbation methods [12,21,27,169,176-179]. Most authors report single-component isotherm results. Multiple-component isotherm data are more complicated because all components are simultaneously in competition for the sorption sites on the stationary phase. The retention time and peak shapes of any solute is dependent on the concentration and properties of all other solutes in the mixture [12,170,180]. For multicomponent mobile phases in liquid and supercritical fluid chromatography this includes each component of the mobile phase.

### 1.5.6 Peak Shape Models

Real chromatographic peaks are rarely truly Gaussian and significant errors can result from the calculation of chromatographic parameters based on this false assumption [153,164,165]. The Gaussian model is only appropriate when the degree of peak asymmetry is slight. Peak asymmetry can arise from a variety of instrumental and chromatographic sources. Those due to extracolumn band broadening (section 1.5.4) and isotherm effects (1.5.5) were discussed earlier. Other chromatographic sources include incomplete resolution of sample components, slow mass transfer processes, chemical reactions, and the formation of column voids [171, 186]. Examples of slow mass transfer processes include diffusion of the solute in microporous solids, polymers, organic gel matrices, and deep pores holding liquid droplets; interactions involving surfaces with a heterogeneous energy distribution; and, in liquid chromatography, interfacial mass transfer resistance caused by poor solvation of bonded phases. Column voids formed by bed shrinkage are usually a gradual process that occurs during the lifetime of all columns and results in progressive peak broadening and/or distortion. A void over the entire cross section of the column near the inlet produces more peak broadening than asymmetry. However, voids occupying only part of the cross section along the length of the bed can produce pronounced tailing or fronting, or even split all peaks into resolved or unresolved doublets. Partial void effects are due to channeling, that is, different residence times in the flow paths formed by the void and packed regions. Slow diffusion in liquids, fails to relax the radial concentration profile fast enough to avoid asymmetry or split peaks. In gas chromatography the phenomenon is less significant because diffusion in gases is much faster.

Meaningful chromatographic data can be extracted from asymmetric peaks by digital integration or curve fitting routines applied to the chromatographic peak profile. The statistical moments of a chromatographic peak in units of time are defined by Eq. (1.37) to (1.39) [153,184,187,188]

Zeroth moment $$M_0 = \int_0^\infty h(t)dt \qquad (1.37)$$

First moment $$M_1 = (1 / M_0)\int_0^\infty t\, h(t)dt \qquad (1.38)$$

Higher moments $$M_n = (1 / M_0)\int_0^\infty (t-M_1)^n\, h(t)dt \qquad (1.39)$$

where h(t) is the peak height at time t after injection. The zeroth moment corresponds to the peak area, the first moment corresponds to the elution time of the center of gravity of the peak (retention time), and the second moment the peak variance. The column plate count is calculated from the first two moments using $N = M_1^2 / M_2$. The third and forth statistical moments measure the peak asymmetry (skew) and the extent of vertical flattening (excess), respectively. For a Gaussian distribution, statistical moments higher than the second have a value of zero. A positive value for the skew indicates a tailing peak. A positive value for the excess indicates a sharpening of the peak profile relative

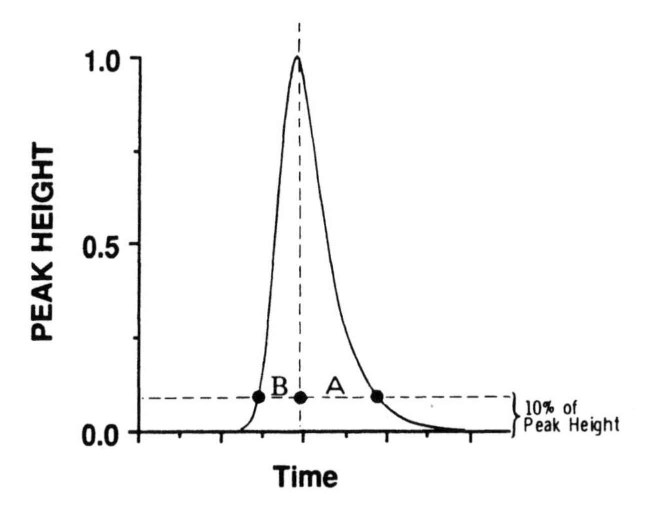


Figure 1.12. The 10% peak height definition of the asymmetry factor. The asymmetry factor is equal to the ratio A / B.

to a Gaussian peak, while a negative value indicates a relative flattening of the upper portion of the peak profile.

Direct numerical integration of the peak profile may lead to many errors and uncertainties arising from the limits used in the integration, baseline drift, noise, and extracolumn dispersion. A slight error in determining the baseline will greatly influence the selected positions for the start and end of the peak resulting in a comparatively large error, particularly for the higher moments. To eliminate these inconsistencies curve fitting of peak profiles by computer or manual methods have been explored [183,184,188-195]. This has led to the general acceptance of the exponentially modified Gaussian (EMG) and exponential Gaussian hybrid functions [194,195] as acceptable models for tailing peaks. The (EMG) is obtained by the convolution of a Gaussian function and an exponential decay function that provides for the asymmetry in the peak profile. The EMG function is defined by three parameters: the retention time and standard deviation of the parent Gaussian function and the time constant of the exponential decay function. The convolution of different functions for the peak front and tail [192,193] and method of data processing [188] increase the flexibility and improve precision, respectively. The column plate count for tailed peaks, $N_{sys}$, can be estimated from the chromatogram, Eq. (1.40), based on the properties of the EMG function for peaks with $1.1 < (A / B) < 2.76$.

$$N_{sys} = [41.7\,(t_R / w_{0.1})^2] / [(A / B) + 1.25] \qquad (1.40)$$

The width at 10% of the peak height ($w_{0.1} = A + B$) and the asymmetry function ($A / B$) are defined as indicated in Figure 1.12. The percent relative error between Eq. (1.40) and the EMG function was < 2%. Another use of the EMG function is to indicate the

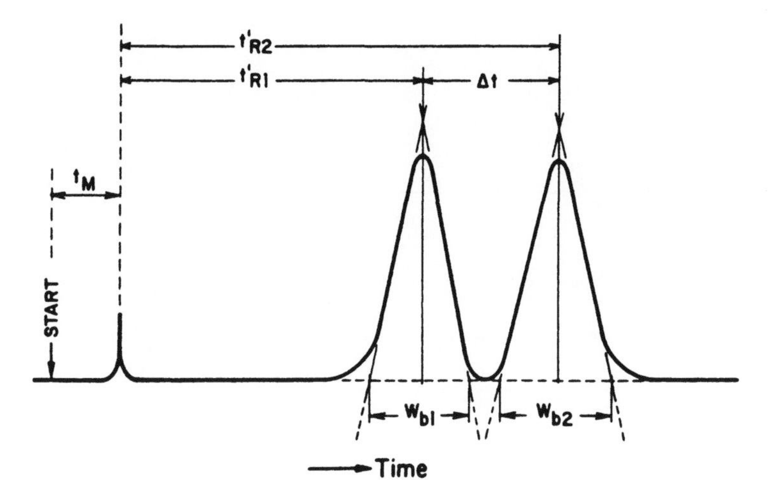


Figure 1.13. Illustration of parameters used to measure the resolution between two neighboring peaks in a chromatogram using Eq. (1.41).

magnitude of extracolumn dispersion assuming that the column behaves as a Gaussian operator [196,197].

## 1.6 RESOLUTION

The separation factor ($\alpha$) is a useful measure of relative peak position in the chromatogram (section 1.4). This function, however, is not adequate to describe peak separations since it does not contain any information about peak widths. The separation of two peaks in a chromatogram is defined by their resolution, $R_S$, the ratio between the separation of the two peak maxima ($\Delta t$) and their average width at base, Figure 1.13, and Eq. (1.41)

$$R_S = 2\Delta t / (w_{b1} + w_{b2}) \quad (1.41)$$

This equation is correct for symmetrical peaks and is easily transposed to use the peak width at half height ($w_h$) when the peak width at base cannot be evaluated for overlapping peaks [$R_S = 1.18\Delta t / (w_{h1} + w_{h2})$]. For two peaks of similar height $R_S = 1.0$ corresponds to a valley separation of about 94% and is generally considered an adequate goal for an optimized separation. Baseline resolution requires $R_S \approx 1.5$. For symmetrical peaks of unequal height and asymmetric peaks a larger value of $R_S$ is required for an acceptable separation as illustrated by the peak ratios in Figure 1.14 [198]. For relatively simple mixtures it is not unusual to set a higher minimum resolution requirement (e.g., $R_S > 2$) in method development to ensure ruggedness as column properties deteriorate.

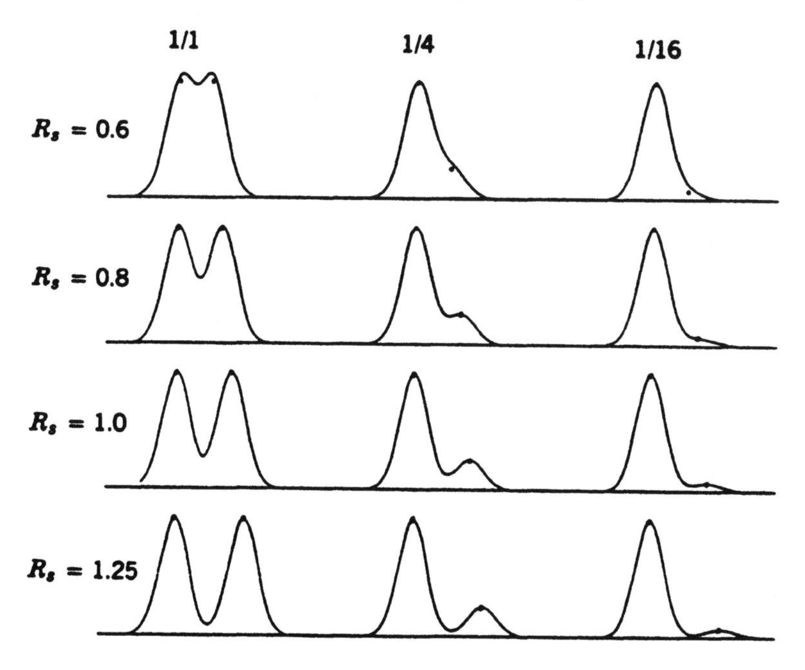


Figure 1.14. Standard resolution curves for the separation of two neighboring peaks as a function of resolution and relative peak area. (From ref [198]; ©John Wiley & Sons)

### 1.6.1 Relationship to Column Properties

A number of equations have been evoked to relate the resolution of two neighboring symmetrical peaks to the adjustable chromatographic variables of selectivity, efficiency, and time [198-202]. The most accurate of these assuming a similar value of N for the two peaks is

$$R_S = [\sqrt{N} / 2][(\alpha - 1) / (\alpha + 1)][k_{AV} / (1 + k_{AV})] \tag{1.42}$$

where $\alpha$ is the separation factor and $k_{AV} = (k_1 + k_2) / 2$ where $k_1$ and $k_2$ are the retention factors of the earlier and later eluting peaks, respectively. For computer-aided method development Eq. (1.42) can be rewritten as Eq. (1.43) by replacing $\alpha$ and $k_{AV}$ with the appropriate individual retention factors

$$R_S = [\sqrt{N} / 2][(k_2 - k_1) / (2 + k_1 + k_2)] \tag{1.43}$$

Probably the most widely quoted of the resolution equations is Eq. (1.44) derived with the assumption that the average of the two peak widths is identical to the peak width of the second peak

$$R_S = [\sqrt{N} / 4][(\alpha - 1) / \alpha][k_2 / (1 + k_2)] \tag{1.44}$$

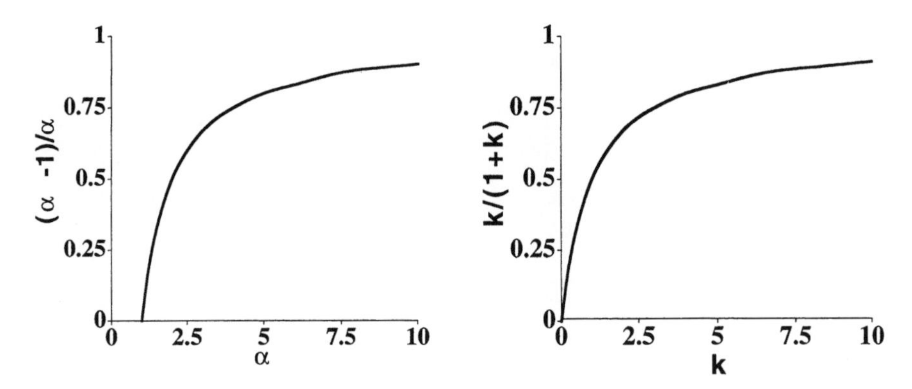


Figure 1.15. Influence of the separation factor and retention factor on the observed resolution for two closely spaced peaks.

Eq. (1.44) should be viewed as a special case of Eq. (1.42) and will tend to underestimate $R_S$ when N is small (< 20,000) and will be increasingly inaccurate for large values of $\Delta t$.

To a first approximation the three terms in Eqs. (1.42) and (1.44) can be treated as independent variables. Eq. (1.44) is sufficiently sound to demonstrate the influence of the column properties on resolution and their optimization. The influence of the separation factor and retention factor on the observed resolution is illustrated in Figure 1.15 for a fixed value of N. For a separation factor of 1.0 there is no possibility of a separation. The separation factor is determined by the distribution constants for the solutes, and in the absence of some difference in distribution constants, and therefore $\alpha$, there will be no separation. Increasing the value of $\alpha$ initially causes a large change in resolution that levels off for large values of $\alpha$. For $\alpha > 2$ separations are easy. Since the separation factor has a large effect on the ease of achieving a certain resolution it is important to choose a chromatographic system that maximizes the separation factor so that the separation can be achieved with the minimum value of N and/or the shortest separation time.

Also from Figure 1.15 it can be seen that resolution is impossible without retention. Initially resolution increases rapidly with retention for $k > 0$. By the time the retention factor reaches a value of around 5, further increases in retention result in only small changes in resolution. The optimum resolution range for most separations occurs for k between 2 and 10. Higher values of k result in excessive separation time with little concomitant improvement in resolution. On the other hand, large values of k do not result in diminished resolution and may be required by necessity for the separation of multicomponent mixtures to accommodate all of the sample components in the separation.

The kinetic properties of columns are the most predictable but the least powerful for optimizing resolution. Resolution only increases with the $\sqrt{N}$. Increasing the column length four- fold will only double resolution at the expense of a four-fold increase in

Table 1.10
Plate count required for $R_S = 1$ for different separation conditions based on Eq. (1.44)

| Retention factor | Separation factor | $N_{req}$ | Retention factor | Separation factor | $N_{req}$ |
|---|---|---|---|---|---|
| 3 | 1.005 | 1,150,000 | 0.1 | 1.05 | 853,780 |
| 3 | 1.01 | 290,000 | 0.2 | 1.05 | 254,020 |
| 3 | 1.02 | 74,000 | 0.5 | 1.05 | 63,500 |
| 3 | 1.05 | 12,500 | 1.0 | 1.05 | 28,200 |
| 3 | 1.10 | 3,400 | 2.0 | 1.05 | 15,800 |
| 3 | 1.20 | 1,020 | 5.0 | 1.05 | 10,160 |
| 3 | 1.50 | 260 | 10 | 1.05 | 8,540 |
| 3 | 2.00 | 110 | 20 | 1.05 | 7,780 |

separation time and an increase in the column pressure drop. In liquid and supercritical fluid chromatography even a four-fold increase in N may be difficult to achieve if standard column configurations are used. Reducing the column diameter in open tubular column gas and supercritical fluid chromatography and the particle size in packed column liquid and supercritical fluid chromatography at a constant column length is a more effective strategy for increasing resolution than increasing the column length. This is somewhat restricted by the increased column pressure drop for liquid and supercritical fluid chromatography while there is greater flexibility in gas chromatography, where this is a more practical strategy.

Equation (1.44) is simply rearranged to predict the plate number required, $N_{req}$, to give a certain separation, Eq. (1.45).

$$N_{req} = 16\, R_S^2[\alpha / (\alpha - 1)]^2[k_2 / (1 + k_2)]^2 \tag{1.45}$$

The plate count required for a resolution of 1.0 using different separation conditions is summarized in Table 1.10 [146]. Practically all chromatographic separations have to be made in the efficiency range of $10^3$-$10^6$ theoretical plates. The importance of optimizing the separation factor and retention factor to obtain an easy separation is obvious from the data in Table 1.10. Easy separations require chromatographic systems that maximize the separation factor and provide at least a minimum value for the retention factor. A common optimization strategy for difficult separations with a limited number of components is to fix the value of the retention factor between 1 and 3 for the two components most difficult to separate in the mixture.

### 1.6.2 Objective Functions

Objective functions are used primarily for computer ranking of separation quality in approaches for automated method development. For this purpose the separation quality throughout the chromatogram must be expressed by a single-valued and easily calculated mathematical function [202-209]. This has proven to be a difficult problem and no universal solution has emerged. There is no straightforward manner to uniquely define the resolution of all peaks simultaneously by a single number. More likely, for any given value of the objective function there will be a large number of separations that could produce the same numerical value, not all of which will agree with the stated goals of the separation. In addition, the objective function may have to consider the number of peaks identified in the chromatogram and the separation time in expressing the goals of the separation. A few representative objective functions and their relative merits are discussed below.

A simple objective function would consider only the separation between the worst separated peak pair, ignoring all others. If a set of chromatograms is to be compared, then this is a reasonable approach, but it does not provide a suitable criterion for locating a single global optimum, since different peaks may show the lowest separation in adjacent chromatograms and many optima will be indicated. The sum of all the resolutions will reflect gradual improvements in different separations, but on its own is of little value. Two peaks that are well resolved and easy to separate will dominate the sum and optimization may result in this peak pair being over separated, while the most difficult pair to separate is ignored. One way of avoiding this problem is to sum the resolutions but to limit the maximum resolution that can be assigned to any peak pair. Taking the separation time into account the chromatographic optimization function (COF) can be defined as

$$\mathrm{COF} = \Sigma_{i=1}^{n} A_i \ln (R_i / R_{id}) + B(t_x - t_n) \tag{1.46}$$

where $R_i$ is the resolution of the ith pair, $R_{id}$ the desired resolution for the ith pair, $t_x$ the maximum acceptable retention time for the last eluted peak, $t_n$ the observed retention time for the last eluted peak, and $A_i$ and B are arbitrary weighting factors used to indicate which peaks are more important to separate than others, and to allow flexibility in setting the acceptable separation time. The COF function gives a single number that tends towards zero as the optimum separation is reached while poor chromatograms produce large, negative values. The COF, however, makes no allowance for peak crossovers and identical values of the COF can result from chromatograms with different numbers of separated peaks. Rather than add the individual resolution values the product of those values can be used. In this case the aim is to space the peaks evenly throughout the chromatogram, since the lowest value of the resolution has a dominant effect. A simple resolution product may still give a higher assessment to an inferior chromatogram but this can be overcome by using the relative resolution product, where the denominator defines the maximum possible value for the resolution product in the given chromatogram. In addition, the relative resolution product can be modified to include a term that incorporates the importance of the separation time in obtaining the desired separation.

None of the functions considered so far specifically takes into account the number of peaks found in the chromatogram. If the object of the separation is to detect the maximum number of peaks, even if the resolution of individual peaks is poor, then Eq. (1.46) and similar equations will be inadequate. A chromatographic response function, CRF, which takes into account the simultaneous importance of resolution, separation time, and the total number of detectable peaks, can be expressed as follows

$$CRF = \Sigma_i R_i + n^a - b(t_x - t_n) - c(t_0 - t_1) \quad (1.47)$$

where n is the number of peaks observed, $t_0$ the minimum desired retention time for the first detected peak, $t_1$ the observed retention time for the first detected peak, and a, b and c are adjustable weighting factors to change the emphasis of the various contributions to the CRF. Usually the exponent is chosen so large (e.g., a = 2) that the appearance of a new peak raises the criterion significantly. A problem with the CRF and similar multi-term functions is that their numerical values can become dominated by one of the terms in the expression and fail to represent the intended mix of terms.

From the above discussion it should be obvious that the selection of an appropriate objective function is a difficult task. It is highly likely that different objective functions will result in the location of different optimum experimental conditions for the separation. Yet, it is not possible to set hard guidelines for the selection of a particular objective function, which must be chosen by practical experience keeping the goals of the separation in mind.

### 1.6.3 Peak Capacity

In the separation of complex mixtures the total number of observed peaks is as important or more so than the resolution of specific peak pairs. Many of these peaks may not be singlets (see section 1.6.4), but systems that separate the mixture into the largest number of observable peaks are desirable unless only a few components are of interest. The separating power of a column can be characterized by its peak capacity, defined as the maximum number of peaks that can be separated with a specified resolution in a given time interval, Figure 1.16. For the general case it can be calculated using Eq. (1.48)

$$n_C = 1 + \int_{t_M}^{t_R} (\sqrt{N} / 4t)dt \quad (1.48)$$

where $n_C$ is the peak capacity and t the separation time [210-214]. For simplicity a resolution of 1.0 is usually adopted and it is assumed that all the peaks are Gaussian. However, the integration of Eq. (1.48) is not straightforward. In many chromatographic systems the plate number is not independent of retention time. Ignoring the variation of the plate number with retention time Eq. (1.48) can be integrated to give

$$n_C = 1 + (\sqrt{N} / 4) \ln (t_R / t_M) \quad (1.49)$$

where $t_M$ is the column hold-up time and $t_R$ the maximum retention time for elution of the last peak. Alternatively, the ratio $(t_R / t_M)$ can be replaced by the ratio $(V_{max} / V_{min})$ where $V_{max}$ and $V_{min}$ are the largest and smallest volumes, respectively, in which a solute can be eluted and detected. The numerical value of $n_C$ depends on $t_R$ and unless

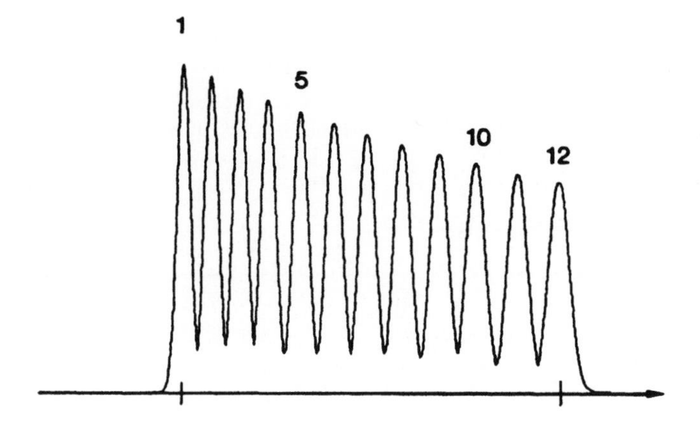


Figure 1.16. Simulation diagram of the column peak capacity.

some method is used to define the useful retention space then unrealistic values for the peak capacity are possible with no connection to chromatographic practice. Using the ratio of ($V_{max}$ / $V_{min}$) is one way to achieve this, although it should be noted that in this form the peak capacity depends on detector characteristics as well as column properties. Eq. (1.49) is a reasonable approximation for packed columns in gas and liquid chromatography but is generally unsuitable for open tubular columns. The relationship between N and the retention factor is more complex for open tubular columns, particularly for small values of the retention factor, for which Eq. (1.49) underestimates $n_C$. Unfortunately, general solutions to Eq. (1.49) that accommodate small retention factors are complex and may require numerical integration or approximations for evaluation [211]. Similar problems arise for programmed separation conditions, for which the best solution is to express the time variation of peak widths by some empirical experimental function for the whole or segments of the separation [213]. The segment values are then summed to provide the total peak capacity. Alternatively, automated measurements of peak widths throughout a chromatogram enable the peak capacity to be calculated by integration of the curve generated from a plot of reciprocal peak widths against retention time [215]. This method has the advantage of providing a rigorous experimental value for the peak capacity. Martin states that typical peak capacities for isocratic liquid chromatography are between 50-70; for programmed liquid chromatographic separations 200-500; programmed separations by gas chromatography up to 2000-5000; and for capillary electrophoretic techniques 500-5000 [216]. The peak capacity for thin-layer (section 6.3.8) and size-exclusion (section 4.9.3.1) chromatography are discussed elsewhere.

The separation number (or Trennzal) is a special case of the peak capacity. It is widely used as a column quality test in gas chromatography [217-219] and to a limited extent in micellar electrokinetic chromatography [220,221]. It is defined as the number of peaks that can be placed between the peaks of two consecutive homologous standards with z and z+1 carbon atoms and separated with a resolution of $R_S$ = 1.177. It represents the integral solution of Eq. (1.49) when the time limits correspond to the retention time of the two homologous standards, usually n-alkanes or fatty acid methyl esters in gas chromatography and alkylaryl ketones in micellar electrokinetic chromatography. Since the peak resolution used to define the peak capacity and the separation number (SN) differ the two terms are related by $n_C = 1.18\ (SN + 1)$, at least for well retained peaks. The separation number is calculated from the retention time ($t_R$) and peak widths at half height ($w_h$) for the homologous standards using Eq. (1.50)

$$SN = ([t_{R(z+1)} - t_{R(z)}] / [w_{h(z)} + w_{h(z+1)}]) - 1 \qquad (1.50)$$

The separation number depends on the properties of the stationary phase, the column length, column temperature and carrier gas flow rate. For meaningful column comparisons it is necessary to standardize the separation conditions. The critical parameters for obtaining reproducible SN values in the temperature program mode are the identity of the standards, the carrier gas flow rate and the temperature program rate [218]. Blumberg and Klee [222] introduced the separation measure, S, as an indication of the separation capacity for isothermal and temperature-programmed gas chromatography. It is evaluated from a series of relationships that depend on the experimental conditions. It has the useful property of being an additive function but has so far not proven popular.

The confident analysis of moderate to complex mixtures requires a large peak capacity. Multidimensional chromatographic techniques that combine distinctly different separation mechanisms for each component dimension provide a powerful approach to obtaining a large peak capacity [223-225]. The peak capacity for a multidimensional system is approximately multiplicative, while series coupling of identical columns only results in an increase in peak capacity that is roughly equal to the product of the square root of the number of coupled columns and their individual peak capacities. Thus, if two identical columns with a peak capacity of 100 were coupled in series, then the resultant peak capacity would be about 140, compared to a value of about 10,000 for two columns of the same peak capacity used in the multidimensional mode. The multiplicative rule provides only an estimate of the peak capacity in two-dimensional separations since correlation of solute retention in the two dimensions reduces the available retention space. Many two-dimensional separation systems employ retention mechanisms that differ in intensity only, for example, the use of reversed-phase liquid chromatography with two different stationary or mobile phases. Consequently the observed total peak capacity is only a fraction of the product of the two one-dimension peak capacities. Coupling of different separation modes, such as reversed-phase liquid chromatography and capillary electrophoresis or normal-phase chromatography and gas chromatography, are examples of two-dimensional systems that can provide large peak capacities. Although a large peak capacity is required for the separation of complex mixtures the actual number of components that can be isolated in a separation is far less than the theoretical peak capacity due to statistical overlap of component peaks. The statistical theory of peak overlap predicts that the ability to resolve peaks in two-dimensional separations does not increase in direct proportion to the increase in peak capacity [226].

### 1.6.4 Statistical Overlap Models

The peak capacity provides an overoptimistic assessment of the real resolving power of a chromatographic system. Real samples do not contain peaks that will emerge exactly at the correct retention time to fulfill the condition of unit resolution. The peak capacity concept can be combined with statistical models that assume that the component peaks of complex mixtures distribute themselves randomly, or according to some other selected probability function, along the retention axis and then solved to indicate the extent of peak overlap [183,226-228]. General theories of statistical overlap are of two types. In the first and more common, the interval between adjacent peaks are interpreted using point-process [229-231] or pulse-point [232] statistics. These models differ in that multicomponent chromatograms are processed as consisting of random point positions in the point-process models and of random pulses at random positions in the pulse-point model. The second approach is based on Fourier analysis to determine the frequency content of the separation [227,233,234]. The correct use of these theories requires a distribution model (such as the Poisson function for randomly distributed peaks). A probability function for the resolution required to establish a minimum peak separation (because the range of peak sizes in the mixture and the degree of saturation affect this). Some model or assumptions concerning the distribution of peak widths in the chromatogram (in programmed separations it can be reasonably assumed that all peaks have the same width). Not surprisingly these models are mathematically complex, but reasonably successful for interpreting the complexity of real chromatograms. Roughly summarized, statistical models of peak overlap suggest that there is a reasonable probability of separating mixtures containing up to about 40 components into single peaks by state-of-the-art chromatographic systems provided that these systems have a large peak capacity. For more complex mixtures or systems with low or moderate peak capacities an increasing number of observable peaks will contain different numbers of unresolved components. Statistical overlap models provide an excellent tool to estimate the number of single component peaks and the number and composition of multiplets of different order. At high saturation (ratio of observed peaks to the peak capacity) the number of single component peaks will be small and most observed peaks will be multiplets of different order.

## 1.7 SEPARATION TIME

In the general theory of chromatography, efficiency and peak capacity are optimized to provide a specified separation in a reasonable time. In certain circumstances it might be more desirable to minimize the separation time and accept some compromise in the separation quality. For example, when a high sample throughput is required or for monitoring changes in fast chemical processes. The separation time is often referred to as the analysis time and time-optimized chromatographic separations as high-speed or fast chromatography [138,154,235,236]. The experimental conditions for time-optimized chromatography can be derived from a simple three-component model. This model assumes that the optimum column length is dictated by the plate number required to resolve the two most difficult to separate components and the total separation time by the time required for the last peak to elute from the column. At a constant temperature (GC) or solvent composition (LC) Eq. (1.45) defines the plate number required for the separation and the corresponding time is given by Eq. (1.3). After some algebraic manipulation, the time required to separate the critical pair is given by Eq. (1.51)

$$t_R = 16R_S^2 \, [\alpha / (\alpha - 1)]^2 \, [(1 + k)^3 / k^2] \, [H / u] \tag{1.51}$$

When the critical pair is separated, so are all other peaks in the chromatogram. If the critical pair is not the last two peaks in the chromatogram the separation time will be $t_R(1 + nk)$ where n is the ratio of the retention factor of the last peak to elute in the chromatogram to the retention factor of the second peak of the critical pair. The fastest separation is obtained when $t_R$ is minimized. This will be the case if the following criteria are met. The minimum useful value for the resolution of the critical pair is accepted. The separation system is optimized to maximize the separation factor ($\alpha$) for the critical pair. The retention factor for the critical pair is minimized ($k \approx 1 - 5$). The column is operated at the minimum value of the plate height, $H_{min}$, corresponding to $u_{opt}$.

For gas chromatography the ratio (H / u) is a complex function of the operating conditions (pressure drop, film thickness, column dimensions, mass transfer properties, etc). Comprehensive theory is available for open tubular and packed columns operated under different conditions providing all relevant equations [130,131,237-239]. Extremely fast separations (separations in a few seconds or less) are only possible when the plate number required to separate the critical pair is low, because short columns are needed for these separations. Reduction in the column diameter for open tubular columns or particle size for packed columns is the best general approach to increase the separation speed in gas chromatography. The price paid is an increase in the column pressure drop. Since open tubular columns are more permeable than packed columns they provide faster separations for most operating conditions. The disadvantage of narrow-bore open tubular columns is their low sample capacity, roughly equal to $2000(r_{c}d_{f})$ where $r_c$ is the column radius (mm) and $d_f$ the film thickness ($\mu$m) when the sample capacity is given in nanograms. Thin-film, narrow-bore columns have typical sample capacities of 10-20 ng per component and are easily overload by major sample components. Packed columns provide high sample capacities, virtually independent of particle size, but are really only useful for fast separations of very simple mixtures [240,241]. Low viscosity and favorable solute diffusion properties make hydrogen the preferred carrier gas. Hydrogen allows higher flow rates or longer columns to be used with a fixed inlet pressure. Helium and nitrogen are 50% and 250%, respectively, slower than hydrogen. Vacuum-outlet operation, for example in GC-MS (section 9.2.2.1), allows faster separations for short and wide-bore columns but is less influential for long or narrow-bore columns (in this case the inlet and average column pressure are hardly influenced by the outlet pressure) [242-245]. Vacuum-outlet operation of short and wide-bore columns allows the entire column to operate at a lower pressure shifting the optimum mobile phase velocity to higher values and providing more favorable solute diffusion coefficients.

High-speed separations, particularly with open tubular columns of 100-μm internal diameter or less, place special demands upon instrumentation [246-250]. The gas chromatograph must be capable of high-pressure operation; sample introduction systems must be capable of delivering extremely small injection band widths (few milliseconds); the detector and data acquisition system must have a fast response (response times of a few milliseconds); and the system should have an extremely small extracolumn dead volume. In addition, the program rate has to be increased in proportion to the reduction in column diameter for temperature programmed separations, requiring ovens that allow very high programming rates (e.g., 100°C/min). Standard gas chromatographs were not designed to meet these specifications and it is only in the last few years that suitable instruments for fast gas chromatography became available.

Liquid chromatographic separations will never be as fast as gas chromatographic separations because mass transfer properties in liquids are inferior to those in gases [138,154,235,236,251]. Most fast separations in liquid chromatography are accomplished at the maximum available inlet pressure. Adopting reduced parameters (section 1.5.3) the separation time is given by

$$t_R = h^2 \phi N_{req}{}^2 \eta / \Delta P \qquad (1.52)$$

The fastest possible separation with a given column pressure drop ($\Delta P$) and $N_{req}$ is obtained at the minimum reduced plate height ($h_{min}$) corresponding to the optimum reduced mobile phase velocity ($\nu_{opt}$). This will be achieved at a specified optimum particle size ($dp_{opt}$) according to

$$dp_{opt} = \sqrt{(h \nu N_{req} D_M \eta \phi / \Delta P)} \qquad (1.53)$$

Since sorbents are available in a limited number of particle sizes, a compromise is necessary. An available particle size is selected that is larger but closest to $dp_{opt}$, The column is made proportionately longer and operated at a slightly higher velocity than $\nu_{opt}$. The retention time, however, will be slightly longer than the theoretical minimum. Elevated temperatures are useful for fast separations because of decreased mobile phase viscosity ($\eta$) and increased solute diffusion ($D_M$).

An alternative strategy for fast liquid chromatography uses short columns packed with small particles operated at high flow rates and often elevated temperatures to separate simple mixtures under conditions were resolution is compromised but still adequate for identification purposes [252-258]. Small diameter particles provide larger plate numbers by virtue of their relatively small interparticle mass transfer resistance combined with a shallow increase in the reduced plate height as the reduced mobile phase velocity is increased beyond its optimum value (see Figure 1.9). Both features are favorable for fast separations. Low column permeability and the limited inlet pressure dictate that short columns are essential, as well as favorable for fast separations, but total plate numbers are low. These short columns (1 – 5 cm) usually packed with 3-μm particles and operated with fast gradients (1-5 min) have found a niche in the pharmaceutical industry for rapid screening of combinatorial libraries and drug metabolite profiling [257,258]. Mass spectrometric detection is frequently used to track analytes to accommodate poor peak separations. It is the synergy between mass separation and chromatographic separation that provides acceptable identification of analytes with separation times typically between 1 to 10 minutes per sample. These miniaturized separation systems require adapted instrumentation because of the small extracolumn volumes and narrow peak profiles.

The fast separation of macromolecules requires special nonporous or perfusive stationary phases with favorable retention and mass transfer properties [138, 259]. Combined with elevated temperatures, steep gradients, high flow velocities and short columns these stationary phases are very effective for fast separations of biopolymers. Supercritical fluids have more favorable kinetic properties for fast separations than liquids but are restricted to operating conditions where the density drop along the column is minimal to avoid unfavorable increases in solute retention with migration distance [260]. Compared with gas chromatography narrow-bore open tubular columns are a poor compromise for fast separations using supercritical fluids because they yield small retention factors and provide low efficiency per unit time. Packed columns with a specific ratio of length to particle diameter to provide the desired resolution are generally used.

Separation speed in capillary electrophoresis is governed by a different set of dynamics compared to liquid chromatography. Efficiency is independent of the column length but migration time is proportional to the length squared [236,261,262]. It is the voltage applied across the capillary column that determines the plate number. Short columns can provide both high plate numbers and fast separations with the minimum column length established indirectly by extracolumn contributions to band broadening and the capacity of the capillary to dissipate the heat generated by the current passing through the column. The migration time ($t_l$) is given by $Ll / \mu V$ where V is the voltage applied across the column of length L, $l$ is the distance migrated by the sample from the point of injection to the point of detection (generally less than L when on-column detection is used), and μ is the mobility of the ion. Separations on a millisecond time scale are possible, if far from routine, using very short capillaries and microstructures (section 8.9.7).

## 1.8 PRINCIPLES OF QUANTIFICATION

This section reviews the basic performance characteristics of chromatographic detectors and the various methods of obtaining quantitative information from their signals.

### 1.8.1 Signal Characteristics

The fundamental properties of the detector signal of general interest are sensitivity, limit of detection, dynamic and linear ranges, response time and noise characteristics. It is convenient to divide chromatographic detectors into two groups based on their response characteristics. Concentration sensitive detectors are non-destructive and respond to a change in mass per unit volume (g/ml). Mass sensitive detectors are destructive and respond to a change in mass per unit time (g/s). Many liquid phase detectors, such as UV-visible, fluorescence, refractive index, etc., are concentration sensitive detectors, while many of the common gas-phase detectors (e.g. flame-based detectors) are mass sensitive. Sensitivity is defined as the detector response per unit mass or concentration of test substance in the mobile phase and is determined as the slope of the calibration curve for detectors with a linear response mechanism. For a concentration sensitive detector the sensitivity, S, is given by $S = AF / w$ and for a mass sensitive detector by $S = A / w$, where A is the peak area, F the detector flow rate, and w the sample amount. A detector with a high sensitivity, corresponding to a larger slope, is better able to discriminate between small differences in the amount of analyte. Sensitivity is often incorrectly used for limit of detection (or limit of determination). Colloquially a detector with a low limit of detection is sometimes incorrectly referred to as a sensitive detector when inferring that the detector provides a useful response to a small amount of sample. A sensitive detector would be one that is able to distinguish a small range of sample sizes. The limit of detection is defined as the concentration or mass flow of a test substance that gives a detector signal equal to some multiple of the detector noise. A value of 2 or 3 is commonly used as the multiple. The limit of detection (D or LOD) is given by $D = aN / S$, where a is the multiple assumed in the definition of the limit of detection. When the test substance is also specified it can be used to compare the operating characteristics of different detectors under standard conditions.

The detector output contains signal associated with the response of the detector to the analyte and noise originating from the interaction of the detector with its environment and from its electronic circuitry [183,263,264]. There are three characteristic types of noise recognized as short term, long term and drift with properties that can change depending on whether they are determined under static or dynamic conditions. Static noise represents the stability of the detector when isolated from the chromatograph. Dynamic noise pertains to the normal operating conditions of the detector with a flowing mobile phase. Ideally, the static and dynamic noise should be similar, and if not detector performance is being degraded by the other components of the chromatograph. The noise signal is measured over a period of time with the detector set to its maximum usable operating range. The observed noise will be different depending on the recording device because virtually all-normal laboratory recording devices include some form of noise filter. Short-term noise is defined as the maximum amplitude for all random variations of the detector signal of a frequency greater than one cycle per minute, Figure 1.17. It is calculated from the recording device by dividing the detector output into a series of time segments less than one minute in duration and summing the vertical displacement of each segment over a fixed time interval, usually 10 to 15 minutes.

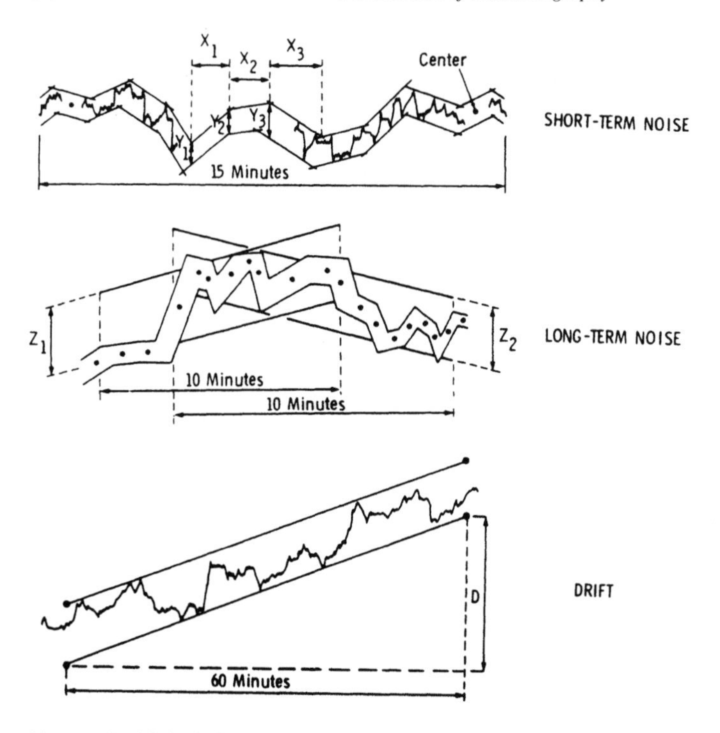


Figure 1.17. Methods for calculating noise for chromatographic detectors.

Long-term noise is the maximum detector response for all random variations of the detector signal of frequencies between 6 and 60 cycles per hour. The long-term noise is represented by the greater of $Z_1$ and $Z_2$ in Figure 1.17. The vertical distances $Z_1$ and $Z_2$ are obtained by dividing the noise signal into ten-minute segments and constructing parallel lines transecting the center of gravity of the baseline deflections. Long-term noise represents noise that can be mistaken for a late eluting peak. Drift is the average slope of the noise envelope measured as the vertical displacement of the baseline over a period of 1h. For spectrophotometric detectors, the signal response is proportional to the path length of the cell and noise values are normalized to a path length of 1 cm.

The dynamic range of the detector is determined from a plot of detector response or sensitivity against sample amount (mass or concentration). It represents the range of sample amount for which a change in sample size induces a discernible change in the detector signal. For many, although not all, chromatographic detectors the relationship between response and analyte mass or concentration is linear for a wide range of analyte concentration. It is the extent of this range that is of most interest to analysts. The linear range is commonly used for all quantitative determinations. It is the range of sample amount over which the response of the detector is constant to within 5%. It is usually expressed as the ratio of the highest sample amount determined from the linearity plot to the limit of detection for the same compound (Figure 1.18).

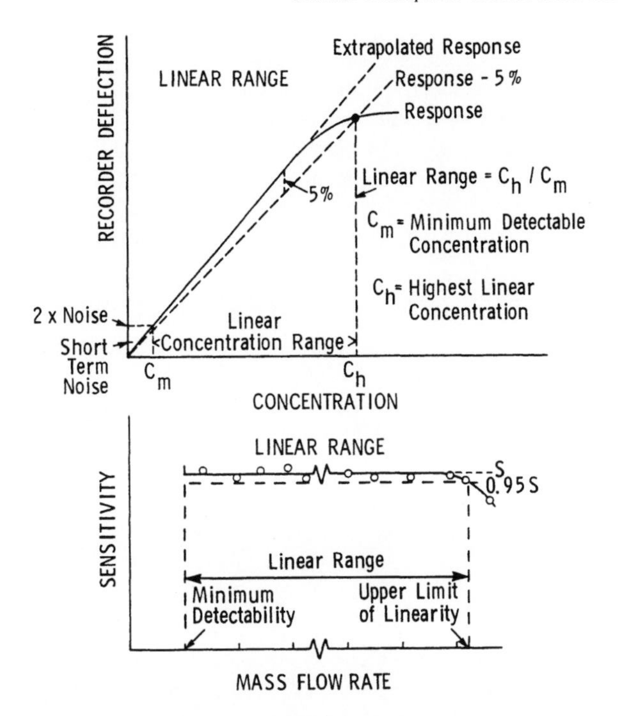


Figure 1.18. Methods for calculating the linear response range for chromatographic detectors.

Table 1.11
Comparison of manual methods for determining peak areas
H = peak height, H' = peak height by construction (see Fig. 1.4) and $w_h$ = peak width at half height, $w_b$ at base, $w_{0.15}$ at 0.15H, $w_{0.25}$ at 0.25H and $w_{0.75}$ at 0.75H

| Method | Calculation | True area (%) | Precision (%) |
|---|---|---|---|
| Gaussian peak | $Hw_h$ | 93.9 | 2.5 |
| Triangulation | $H'w_b/2$ | 96.8 | 4.0 |
| Condol-Bosch | $H(w_{0.15} + w_{0.75})/2$ | 100.4 | 2.0 |
| EMG | $0.753Hw_{0.25}$ | 100 | 2.0 |
| Planimetry | | 100 | 4.0 |
| Cut and weigh | | 100 | 1.7 |

## 1.8.2 Integration Methods

For quantitative analysis it is necessary to establish a relationship between the magnitude of the detector signal and sample amount. The detector signal is measured by the peak area or height from the chromatogram. A number of manual methods are available for calculating peak areas, Table 1.11 [3,184,264,265]. For Gaussian peaks the product of the peak height and the width at half height or the method of triangulation can be used. In both cases the calculated area is less than the true peak area. This is only important if the absolute peak area is used in further calculations. In this case the calculated area should be multiplied by the appropriate factor to give the true peak area. When peak area ratios are used, as in calibration, this is not important. The Condol-Bosch and exponentially modified Gaussian (EMG) models provide a reasonable estimate of the peak area for Gaussian and moderately asymmetric peaks. Up to an asymmetry factor of 3 in the case of the EMG method. No single method is perfect and common problems include the difficulty of accurately defining peak boundaries, variable precision between analysts and the need for a finite time to make each measurement. A major disadvantage of manual measurements is the necessity that all peaks of interest must be completely contained on the chart paper (or adjusted to remain on the chart paper by varying the detector attenuation during the separation). This severely limits the dynamic range of solute composition that can be analyzed. For those methods that depend on the measurement of peak widths narrow peaks are usually difficult to measure with acceptable accuracy using a magnifying reticule or comparator unless fast chart speeds are used to increase the peak dimensions.

Planimetry and cutting out and weighing of peaks require no assumptions about the shape of the peak profile and can be used to determine the area of asymmetric peaks. The proper use of a planimeter (a mechanical device designed to measure the area of any closed plane by tracing out the periphery of the plane with a pointer connected by an armature to a counter) requires considerable skill in its use. Even so, obtaining accurate results requires repetitive tracing of each peak with the totals averaged. The cut and weigh procedure depends critically on the accuracy of the cutting operation. The homogeneity, moisture content and weight of the paper influence precision. Copying the chromatogram onto heavy bond paper, with expansion if possible, will preserve the original chromatographic record of the separation and enhance the precision of the weighings.

Dedicated electronic integrators and personal computers with appropriate software for integration are routinely used for recording chromatograms [183,184,264,266-271]. Since manual methods for determining peak information are tedious and slow, this is hardly surprising. It is important, however, to understand the limitations of electronic integration. Computer-based systems are increasingly used since they can combine instrument control functions with chromatogram recording and integration as well as providing electronic data storage. Computer-based systems also provide flexible approaches for reporting results and for performing advanced data analysis techniques using additional resident software.

The continuous voltage output from chromatographic detectors is not a suitable signal for computer processing. The conversion of the detector signal to a computer readable form requires an interface usually resident as an expansion card in the computer. The interface scales the detector output to an appropriate range, digitizes the signal, and then transfers the data to a known location in the computer. The original input signal is transformed into a series of voltages on a binary counter and is stored as a series of binary words suitable for data processing.

The important characteristics of the analog to digital conversion device are its sampling frequency, resolution and range. The accurate recording of chromatographic peaks requires that at least 10 data points are collected over the peak width. The sampling frequency must match the requirements for the narrowest peak in the chromatogram. For modern devices with sampling frequencies of 5 Hz or better undersampling is usually only a problem in high-speed chromatography [250,268]. Analog to digital converters used for chromatographic applications are usually auto-ranging, meaning that the detector signal is automatically divided into ranges so as to provide sufficient resolution near the baseline for peak detection while accurately registering the peak maximum for large signals. The resolution of the analog to digital converter is the smallest change in the analog signal that can be seen in the digital output (specified as the number of bits).

Converted data is usually averaged (bunched) to minimize storage space. Using the local peak width to determine the frequency of bunching creates uniform sampling density throughout the chromatogram. Long stretches of stable baseline are stored in one bunch represented by a single datum and the number of times it recurs. Stored data is initially smoothed and peak locations identified by a slope sensitivity function. Small peaks and baseline artifacts are removed using a threshold function or later by a minimum area reject function. Standard procedures may fail in the case of complex baselines, tailing peaks or excessive peak overlap. Computer-based integration software often tries to compensate for this problem by providing the possibility of video integration. The operator can set any integration boundaries desired by moving a cursor on the video monitor. Choosing boundaries though is arbitrary, and it is quite likely that different operators will choose different boundaries. Because this procedure can not be validated it is unsuitable for regulatory and general quality assurance problems. It can be useful in other circumstances to correct blunders made by the software in logically interpreting the correct boundary positions.

Most often, the derivatives of the smoothed signal are used for peak detection [183,184,270-275]. Peaks are detected because the detector signal amplitude changes more rapidly when peaks elute than the baseline signal does between peaks. There is a threshold of slope below which peaks are not detected as different to baseline fluctuations, and this value is set by the slope sensitivity factor. Differentiation of the detector signal enhances the changes within the signal facilitating the accurate location of peak start and end positions. The starting point of the peak is detected when the first derivative has reached a predetermined threshold value. The peak maximum and also the valley between partially resolved peaks are then indicated where the derivative falls to zero and the end of the peak where the derivative drops below the threshold and the signal returns to the baseline. In order to avoid the detection of false peaks due to abrupt changes caused by baseline noise a minimum peak width is usually predefined and the detected peak is accepted only if the difference between the start and end of the peak exceeds the threshold. Some systems use a two step peak-indicating algorithm. In the first step a coarse estimation of peak positions is made when the first derivative exceeds the threshold for two or more consecutive data points. Then the second derivative of the signal is analyzed backwards and peak positions reassigned with respect to the chosen threshold in the usual way. The choice of threshold value is very important and is often user selectable. Alternatively, the integrator may use the average value of the signal fluctuations at the start of the chromatogram to automatically self-program an appropriate threshold value. With a low threshold, noise cannot be distinguished from peaks, and too many signals will be reported. With too high a threshold, small peaks are overlooked and there will be a late start to the integration of detected peaks.

The area between the start and end positions identified for the peak is integrated using a summation algorithm usually based on Simpson's rule or trapezoidal integration. Simpson's rule is more accurate because it fits a quadratic function to groups of three consecutive data points whereas the trapezoid method involves fitting a straight line between data points. An adequate sampling frequency is also important since this determines the number of slices across the peak (i.e. the data points that are integrated). Subtracting the baseline area from the accumulated integral count completes the peak area calculation.

In the absence of significant baseline noise most integrators should be capable of high precision and accuracy when integrating isolated symmetrical peaks. Examples of peaks often poorly treated by computing integrators are small peaks with large peak widths, peaks on the tail of larger peaks or the solvent front, peaks on a noisy baseline and fused peaks [183,184,268,276,277]. Noise blurs the determination of base peak widths by making it difficult to locate the exact peak start and end positions. Noise at the peak tops can cause integrators to split area measurements when valley recognition is triggered by the micro-peaks. Ultimately, noise determines the smallest peak size that can be distinguished from random baseline fluctuations.

Most computing integrators construct baselines as a straight line drawn beneath the peak. Real chromatograms with a rising baseline may have a more complex baseline structure resulting in both gains and losses in the true peak height and area depending on the shape of the real baseline, Figure 1.19. Fused peaks are generally integrated by dropping perpendiculars from the valley between them or, if one peak is much smaller than the other and located on its tail, by skimming a tangent below it, Figure 1.19. The measurement errors introduced by perpendicular separation of two peaks are determined by their relative sizes, their asymmetries, the elution order and degree of overlap [184,270,277-279]. If peaks are symmetrical but unequal in size, perpendicular separation over-estimates the smaller area at the expense of the larger one. If the peaks are asymmetric there is a larger inequality in the contribution of each peak to the other. Tangent skimming under-estimates the area of the small peak unless it is very much smaller and narrower than the peak from which it is skimmed. The general conclusion is that there is no secure method for the accurate integration of fused peaks.

Area is the *de facto* peak characteristic related to the mass or concentration of analyte detected [184,270,278-283]. Height is susceptible to peak asymmetry and retention differences while area is not, and the linear dynamic range for area measurements is greater than for height. Under some circumstances peak height might provide a more reliable or reproducible estimate of the analyte present. When the signal to noise ratio is small, height is preferred because errors of baseline placement affect height less than area. For manual measurements of well-separated and symmetrical peaks

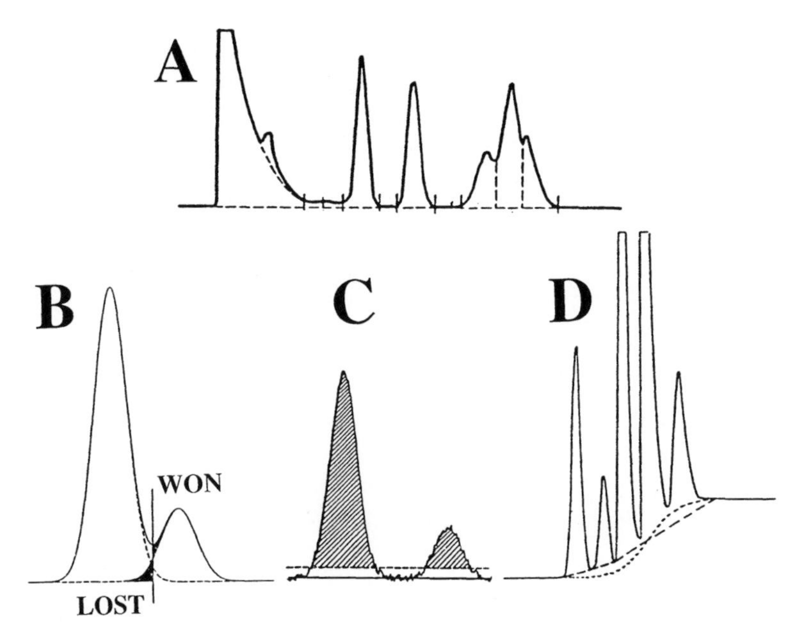


Figure 1.19. Possible error sources in the integration of chromatographic peaks. (A) Tangent skimming and perpendicular drop methods to assign peak boundaries for fused peaks; (B) the misproportionation of the true peak area between two incompletely separated peaks of different size; (C) loss of peak area due to integration with a threshold value set too high; and (D) variation of peak area due to differences between the real sloping baseline and the interpolated integrator baseline.

height is less tedious to determine than area. In other circumstances the precision and accuracy of peak height and area measurements depend on several chromatographic variables, including sample size, mobile phase composition, flow rate, and column temperature. When a mass-sensitive detector is used, quantification should be based on peak area since precision of the measurement will be independent of flow rate, temperature stability and other chromatographic factors that alter either the elution volume or peak shape. In contrast, when a concentration-sensitive detector is used the peak height should be independent of flow rate variations and could be more precise than area measurements. For liquid chromatography area measurements are preferred when the column flow rate can be controlled precisely even if the mobile phase composition shows some variability and *vice versa* as far as peak height measurements are concerned. Quantification in gradient elution chromatography requires careful control of total flow rate when peak areas are used and gradient composition when peak heights are used. To test which alternative is applicable, variation in the retention time of early eluting peaks indicates poor flow precision and variation in the retention time of late eluting peaks suggests poor precision in the mobile phase composition. Consequently, whether peak height or peak area is selected for a particular analysis depends on system performance and not necessarily on sample composition. For fused peaks, peak height is sometimes more accurate for tailing peaks of equal area as well as for the smaller of two peaks when the small peak is eluted first and the area ratio is large and for the smaller of two peaks when the small peak is eluted last if the peaks are symmetrical.

### 1.8.3 Relative Composition

Four techniques are commonly used to convert peak heights or areas into relative composition data for the sample. These are the normalization method, the external standard method, the internal standard method and the method of standard additions [264,284]. In the normalization method the area of all peaks in the chromatogram are summed and then each analyte is expressed as a percentage of the summed areas. All sample components must elute from the column and their responses must fall within the linear operating range of the detector. This method will always lead to totals representing 100%. If the detector response is not the same for all compounds then response factors are required to adjust the peak areas to a common scale. Response factors are usually determined as the slope of the calibration curve and converted to relative response factors since these tend to be more stable than absolute values.

The external standard method compares peak areas in an unknown sample with the areas produced by a known amount of the same analytes in a standard sample analyzed under identical conditions. Standards are chromatographed separately alternating in order with samples for the highest precision. The standard will normally be a synthetic mixture with individual components at a similar concentration to those of the sample. No response factors are required, therefore, and peak shape and linearity problems are avoided. The standard solutions should be prepared in the same matrix as the sample. High precision requires absolute constancy of the separation conditions. Equal size injections are essential and injection accuracy and precision are critical parameters. Chromatographic conditions must be stable and all measurements made within the linear operating range of the detector. External standards are frequently employed in quality control applications of raw materials, drugs and formulations, etc., where mostly the major components are analyzed and strict requirements on accuracy and precision of the method apply (0.5 to 2.0% RSD). Note that if the standard and sample components are not identical or one standard is used to determine more than one component, then the appropriate response factors must be determined as described for the normalization method. Using the deferred standard method can minimize the separation time for the external standard method. Injection of the standard is delayed until some time after the sample injection so that it elutes in a region of the chromatogram free of other components.

There are two distinct uses of an internal standard. It can be a substance or substances added to the sample prior to injection to allow absolute quantities of analytes to be measured by compensating for variation in the injected sample volume. Alternatively, it can be a substance or substances that are added to the sample at the earliest possible point in an analytical scheme to compensate for sample losses occurring during sample preparation and chromatographic analysis. A suitable internal standard must be available in a pure form, have good storage properties and be unreactive towards the sample and its matrix. Since it is added to the sample it cannot be a normal component of the sample and must be completely separated from all other components of the sample. Since it is important that it behaves like the sample components it must have similar physical and chemical properties to them and preferably it should have a similar detector response factor and be present in a similar amount. These conditions are rarely, if ever, met in full, particularly when a single standard is used for the analysis of a multicomponent mixture with sample components varying in relative composition and chemical type. Substances commonly used as internal standards include analogs, homologs, isomers, enantiomers (if a chiral separation system is used), stable isotopically labeled compounds (if a mass detector is used) and radioactive isotopically labeled compounds (if particle detector is used). Analogs and homologs are perhaps the most widely used substances simply because they are likely to be available. Stable isotopically labeled internal standards are frequently used in gas chromatography-mass spectrometry where the mass discriminating power of the mass spectrometer can be used to differentiate between the analyte and internal standard.

When using the internal standard method calibration curves are first prepared for all analytes to be determined and the internal standard to establish the relevant response factors. A constant amount of standard is added to each sample, preferably at a concentration similar to the analytes of interest, and a calibration curve constructed from the ratio of the detector response to the analyte divided by the response to the internal standard plotted against the concentration (amount) of analyte. The ratio of the detector response to the sample unknowns and internal standard is then used for all quantitative measurements. A general problem in the calibration method arises if a least squares fit is used. The error distribution may be no longer normally distributed and when low precision is obtained for the internal standard biased values for the estimated sample concentrations are possible [285]. Poor precision due to variation of the injection volume can largely be eliminated by use of an internal standard. This is frequently a significant error in gas chromatography using manual injection.

The choice of an internal standard for an analytical procedure is often made in a too cavalier a fashion and may actually provide lower precision than external calibration. The successful use of an internal standard depends on the existence of a high correlation between the peak areas (heights) of the analytes and the internal standard for the complete analytical procedure and their being lower variability in the internal standard area (height) compared to those of the analytes [286-288]. An external standard will generally provide higher precision than an internal standard when the variation of the recovery for analytes and standards is sufficiently different and the standard deviation in the mean for repeated analyses of the internal standard is larger than that for the analytes of interest

The method of standard additions is the least used method of quantitation in chromatography because it requires several repetitive separations to yield a single result [284,289-291]. Equal volumes of the sample solution are taken, all but one are spiked with known and different amounts of the analyte, and all are then diluted to the same volume. Each sample is analyzed individually and a pseudocalibration curve produced using the known amounts of analyte as the concentration (weight) axis. Extrapolation of the pseudocalibration curve to zero detector response provides an accurate value for the unknown analyte concentration (amount). A lack of linearity in the pseudocalibration plot is a good indication of matrix effects. Analytes and standards must be identical, all solutions analyzed must fall within the linear range of the detector response, absolute constancy of analytical conditions is required (same as for the external standard method), and injection volumes should be identical. Data may be treated statistically by linear regression techniques if a sufficient number of measurements are made.

## 1.9 REFERENCES


[1] L. S. Ettre and K. I. Sakodynskii, Chromatographia 35 (1993) 329.
[2] L. S. Ettre, Chromatographia 42 (1996) 343.
[3] A. Braithwaite and F. J. Smith, *Chromatographic Methods*, Blackie Academic & Professional, Glasgow, 5 th. Edn., 1996.
[4] L. S. Ettre, Anal. Chem. 47 (1975) 422A.
[5] J. C. Touchstone, J. Liq. Chromatogr. 16 (1993) 1647.
[6] L. S. Ettre and A. Zlatkis (Eds.), *75 Years of Chromatography. A Historical Dialogue*, Elsevier, Amsterdam, 1979.
[7] C. W. Gehrke, R. L. Wixom and E. Bayer (Eds.), *Chromatography – A Century of Discovery 1900-2000. The Bridge to the Sciences/Technology*, Elsevier, Amsterdam, 2001.
[8] H. J. Issaq (Ed.), *A Century of Separation Science*, Dekker, New York, NY, 2001.
[9] L. S. Ettre, Pure & Appl. Chem. 65 (1993) 819.
[10] L. S. Ettre, Chromatographia, 38 (1994) 521.
[11] A. S. Rathore and Cs. Horvath, J. Chromatogr. A 743 (1996) 231.
[12] G. Guiochon and S. G. Shirazi and A. M. Katti, *Fundamentals of Preparative and Nonlinear Chromatography*, Academic Press, London, 1994.
[13] U. Wenzel, J. Chromatogr. A 928 (2001) 1.
[14] L. S. Ettre and J. V. Hinshaw, Chromatographia 43 (1996) 159.
[15] L. M. Blumberg, Chromatographia 44 (1997) 326.
[16] J. F. Parcher, Chromatographia 47 (1998) 570.
[17] V. A. Davankov, L. A. Onuchak, S. Yu. Kudryashov and Yu. I. Arutyunov, Chromatographia 49 (1999) 449.
[18] G. Foti and E. Kovats, J. High Resolut. Chromatogr. 23 (2000) 119.
[19] J. A. G. Dominguez, J. C. Diez-Masa and V. A. Davankov, Pure Appl. Chem. 73 (2001) 969.
[20] V. G. Berezkin and I. V. Malyukova, J. Microcol. Sep. 11 (1999) 125.
[21] J. R. Conder and C. L. Young, *Physicochemical Measurements by Gas Chromatography*, Wiley, New York, 1979.
[22] R. J. Laub and P. L. Pecsok, *Physicochemical Applications of Gas Chromatography*, Wiley, New York, 1978.
[23] V. L. McGuffin and S.-H. Chen, Anal. Chem. 69 (1997) 930.
[24] M. C. Ringo and C. E. Evans, Anal. Chem. 69 (1997) 4964.
[25] J. E. MacNair, K. D. Patel and J. W. Jorgenson, Anal. Chem. 71 (1999) 700.
[26] R. Ohmacht and B. Boros, Chromatographia 51 (2000) S-205.
[27] N. A. Katsanos and F. Roubani-Kalantzopoulou, Adv. Chromatogr. 40 (2000) 231.
[28] N. A. Katsanos, R. Thede and F. Roubani-Kalantzopoulou, J. Chromatogr. A 795 (1998) 133.

[29] R. L. Grob (Ed.), *Modern Practice of Gas Chromatography*, Wiley, New York, 1995.
[30] D. C. Locke, Adv. Chromatogr. 14 (1976) 87.
[31] K. Kojima, S. Zhang and T. Hiaki, Fluid Phase Equilibria 131 (1997) 145.
[32] F. R. Gonzalez, J. Chromatogr. A 942 (2002) 211.
[33] S. K. Poole and C. F. Poole, J. Chromatogr. 500 (1990) 329.
[34] K. G. Furton and C. F. Poole, J. Chromatogr. 399 (1987) 47.
[35] E. F. Meyer, J. Chem. Edu. 50 (1973) 191.
[36] R. C. Castells, J. Chromatogr. 350 (1985) 339.
[37] E. Katz, R. Eksteen, P. Schoenmakers and N. Miller (Eds.), *Handbook of HPLC*, Marcel Dekker, New York, 1998.
[38] T. L. Hafkenscheid and E. Tomlinson, Adv. Chromatogr. 25 (1986) 1.
[39] D. W. Armstrong, Adv. Chromatogr. 39 (1998) 239.
[40] R. Kaliszan, *Quantitative Structure – Chromatographic Retention Relationships*, Wiley, New York, 1987.
[41] C. F. Poole, S. K. Poole and A. D. Gunatilleka, Adv. Chromatogr. 40 (2000) 159.
[42] S. K. Poole and C. F. Poole, J. Chromatogr. 845 (1999) 381.
[43] A. D. Gunatilleka and C. F. Poole, Anal. Commun. 36 (1999) 235.
[44] P. W. Carr, Microchem. J. 48 (1993) 4.
[45] M. H. Abraham and H. S. Chadha in *Lipophilicity in Drug Action and Toxicology*, V. Liska, B. Testa and H. van de Waterbeemd (Eds), VCH, Weinheim, 1996, p. 311-337.
[46] P. Suppan and N. Ghoneim, *Solvatochromism*, The Royal Society of Chemistry, Cambridge, UK, 1997.
[47] A. de Juan, G. Fonrodona and E. Casassas, Trends Anal. Chem. 16 (1997) 52.
[48] M. H. Abraham in *Quantitative Treatment of Solute/Solvent Interactions*, P. Politzer and J. S. Murray (Eds.), Elsevier, Amsterdam, 1994, p. 83-134.
[49] M. H. Abraham, Chem. Soc. Revs. 22 (1993) 73.
[50] M. H. Abraham, C. F. Poole and S. K. Poole, J. Chromatogr. A 842 (1999) 79.
[51] C. F. Poole and S. K. Poole, J. Chromatogr. A 965 (2002) 263.
[52] M. H. Abraham and J. A. Platts, J. Org. Chem. 66 (2001) 3484.
[53] M. H. Abraham, C. M. Du and J. A. Platts, J. Org. Chem. 65 (2000) 7114.
[54] M. H. Abraham and J. C. McGowan, Chromatographia 23 (1987) 243.
[55] M. H. Abraham, P. L. Grellier and R. A. McGill, J. Chem. Soc. Perkin Trans. 2, (1987) 797.
[56] M. H. Abraham and G. S. Whiting, J. Chromatogr. 594 (1992) 229.
[57] M. H. Abraham, J. Chromatogr. 644 (1993) 95.
[58] M. H. Abraham, J. Andovian-Haftvan, G. S. Whiting, A. Leo and R. S. Taft, J. Chem. Soc. Perkin Trans. 2, (1994) 1777.
[59] M. Mutelet and M. Rogalski, J. Chromatogr. A 923 (2001) 153.
[60] Q. Li, C. F. Poole, W. Kiridena and W. W. Koziol, Analyst 125 (2000) 2180.
[61] M. H. Abraham, G. S. Whiting, R. M. Doherty, W. J. Shuely, J. Chem. Soc. Perkin Trans. 2, (1990) 1451.
[62] J. D. Weckwerth and P. W. Carr, Anal. Chem. 70 (1998) 4793.
[63] M. H. Abraham, H. S. Chadha and R. C. Mitchell, J. Pharm. Sci. 83 (1994) 1257.
[64] M. H. Abraham, H. S. Chadha and R. C. Mitchell, J. Pharm. Pharmacol. 47 (1995) 8.
[65] M. H. Abraham, R. Kumarsingh, J. E. Cometto-Muniz, W. S. Cain, M. Roses, E. Bosch and M. L. Diaz, J. Chem. Soc. Perkin Trans. 2, (1998) 2405.
[66] M. H. Abraham, P. L. Grellier, D. V. Prior, P. P. Duce, J. J. Morris and P. J. Taylor, J. Chem. Soc. Perkin Trans 2, (1989) 699.
[67] M. H. Abraham, P. L. Grellier, D. V. Prior, J. J. Morris and P. J. Taylor, J. Chem. Soc. Perkin Trans. 2, (1990) 521.
[68] M. H. Abraham, J. Phys. Org. Chem. 6 (1993) 660.
[69] A. M. Zissimos, M. H. Abraham, M. C. Barker, K. J. Box and K. Y. Tam, J. Chem. Soc. Perkin Trans. 2, (2002) 470.

[70] M. H. Abraham, G. S. Whiting, R. M. Doherty and W. J. Shuely, J. Chromatogr. 587 (1991) 213.
[71] J. A. Platts, D. Butina, M. H. Abraham and A. Hersey, J. Chem. Inf. Comput. Sci. 39 (1999) 835.
[72] C. F. Poole, S. K. Poole and M. H. Abraham, J. Chromatogr. A 798 (1998) 207.
[73] M. D. Trone and M. G. Khaledi, J. Chromatogr. A 886 (2000) 245.
[74] E. Fuguet, C. Rafols, E. Bosch, M. H. Abraham and M. Roses, J. Chromatogr. A 942 (2002) 237.
[75] C. F. Poole, S. K. Poole, D. S. Seibert and C. M. Chapman, J. Chromatogr. B 689 (1997) 245.
[76] C. F. Poole, A. D. Gunatilleka and R. Sethuraman, J. Chromatogr. A 885 (2000) 17.
[77] R. M. Smith, J. Chromatogr. A 656 (1993) 381.
[78] R. M. Smith (Ed), *Retention and Selectivity in Liquid Chromatography*, Elsevier, Amsterdam, 1995.
[79] A. Vailaya and C. Horvath, J. Phys. Chem. B 102 (1998) 701.
[80] F. R. Gonzalez J. L. Alessandrini and A. M. Nardillo, J. Chromatogr. A 810 (1998) 105.
[81] J. Li, Anal. Chim. Acta 369 (1998) 21.
[82] C. M. Bell, L. C. Sander and S. A. Wise, J. Chromatogr. A 757 (1997) 29.
[83] A Peter, G. Torok, D. W. Armstrong, G. Toth and D. Tourue, J. Chromatogr. A 828 (1998) 177.
[84] C. F. Poole and T. O. Kollie, Anal. Chim. Acta 282 (1993) 1.
[85] S. D. Martin, C. F. Poole and M. H. Abraham, J. Chromatogr. A 805 (1998) 217.
[86] G. Sheng, Y. Shen and M. L. Lee, J. Microcol. Sep. 9 (1997) 63.
[87] Y. Mao and P. W. Carr, Anal. Chem. 72 (2000) 110.
[88] P. L. Zhu, J. W. Dolan and L. R. Snyder, J. Chromatogr. A 756 (1996) 41 and 63.
[89] D. Bolliet and C. F. Poole, Analyst 123 (1998) 295.
[90] T. M. Pawlowski and C. F. Poole, Anal. Commun. 36 (1999) 71.
[91] H. Chen and C. Horvath, J. Chromatogr. A 705 (1995) 3.
[92] J. Li, Y. Hu and P. W. Carr, Anal. Chem. 69 (1997) 3884.
[93] D. E. Henderson and D. J. O'Connor, Adv. Chromatogr. 23 (1984) 65.
[94] C. Wolf and W. H. Pirkle, J. Chromatogr. A 785 (1997) 173.
[95] J. Li and P. W. Carr, J. Chromatogr. A 670 (1994) 105.
[96] R. Ranatunga, M. F. Vitha and P. W. Carr, J. Chromatogr. A 946 (2002) 47.
[97] R. R. Krug, Ind. Eng. Chem. Fundam. 19 (1980) 50.
[98] J. D. Thompson, J. S. Brown and P. W. Carr, Anal. Chem. 73 (2001) 3340.
[99] T. Greibrokk and T. Andersen, J. Sep. Sci. 24 (2001) 899.
[100] K. Ryan, N. M. Djordjevic and F. Erni, J. Liq. Chromatogr. & Rel. Technol. 19 (1996) 2089.
[101] M. H. Chen and C. Horvath, J. Chromatogr. 788 (1997) 51.
[102] B. W. Yan, J. H. Zhao, J. S. Brown and P. W. Carr, Anal. Chem. 72 (2000) 1253.
[103] J. C. Giddings, *Dynamics of Chromatography*, Marcel Dekker, New York, 1965.
[104] E. Grushka, L. R. Snyder and J. H. Knox, J. Chromatogr. Sci. 13 (1975) 25.
[105] S. J. Hawkes, J. Chem. Edu. 60 (1983) 393.
[106] J. Cazes and R. P. W. Scott, *Chromatography Theory*, Dekker, New York, 2002.
[107] J. H. Knox, Adv. Chromatogr. 38 (1998) 1.
[108] J. H. Knox, J. Chromatogr. A 831 (1999) 3.
[109] A. J. P. Martin and R. L. M. Synge, Biochem. J. 35 (1941) 1358.
[110] G. Guiochon, J. Chromatogr. Rev. 8 (1966) 1.
[111] C. A. Cramers, J. A. Rijks and C. P. M. Schutjes, Chromatographia 14 (1981) 439.
[112] L. M. Blumberg, J. High Resolut. Chromatogr. 22 (1999) 213.
[113] D. P. Poe and D. E. Martire, J. Chromatogr. 517 (1990) 3.
[114] T. Farkas, G. Zhong and G. Guiocon, J. Chromatogr. A 849 (1999) 35.
[115] W. Jennings, E. Mittlefehldt and P. Stremple. *Analytical Gas Chromatography*, Academic Press, San Diego, CA, 1997.
[116] S. G. Harding and H. Baumann, J. Chromatogr. A 905 (2001) 19.
[117] U. Tallarek, E. Bayer and G. Guiochon, J. Am. Chem. Soc. 120 (1998) 1494.
[118] R. A. Shalliker, B. S. Broyles and G. Guiochon, J. Chromatogr. A 826 (1998) 1.
[119] J. J. Van Deemter, F. J. Zuiderweg and A. Klinkenberg, Chem. Eng. Sci. 50 (1995) 3869.

[120] J. H. Knox and H. P. Scott, J. Chromatogr. 282 (1983) 297.
[121] K. Ogan and R. P. W. Scott, J. High Resolut. Chromatogr. 7 (1984) 382.
[122] P. Sandra, J. High Resolut. Chromatogr. 12 (1989) 273.
[123] V. G. Berezkin, I. V. Malyukova and D. S. Avoce, J. Chromatogr. A 872 (2000) 111.
[124] L. M. Blumberg, J. High Resolut. Chromatogr. 20 (1997) 597.
[125] A. K. Bemgard, L. G. Blomberg and A. L. Colmsjo, Anal. Chem. 61 (1989) 2165.
[126] P. A. Leclerc and C. A. Cramers, J. High Resolut. Chromatogr. 8 (1985) 764.
[127] L. S. Ettre, Chromatographia 17 (1983) 553.
[128] J. H. Knox and M. T. Gilbert, J. Chromatogr. 186 (1979) 405.
[129] R. Swart, J. C. Kraak and H. Poppe, Chromatographia 40 (1995) 587.
[130] R. P. W. Scott, *Introduction to Analytical Gas Chromatography*, Marcel Dekker, New York, 1998.
[131] R. J. Jonker, H. Poppe and J. F. K. Huber, Anal. Chem. 54 (1982) 2447.
[132] M. M. Robson, K. D. Bartle and P. Myers, J. Microcol. Sep. 10 (1998) 115.
[133] C. Horvath and H.-J. Lin, J. Chromatogr. 49 (1987) 43.
[134] Y. L. Yang and M. L. Lee, J. Microcol. Sep. 11 (1999) 131.
[135] E. Katz, K. L. Ogan and R. P. W. Scott, J. Chromatogr. 270 (1983) 51.
[136] K. Miyabe and G. Guiochon, Anal. Chem. 72 (2000) 1475.
[137] Y. Shen, Y. J. Yang and M. L. Lee, Anal. Chem. 69 (1997) 628.
[138] H. Chen and Cs. Horvath, J. Chromatogr. A 705 (1995) 3.
[139] G. Sheng, Y. Shen and M. L. Lee, J. Microcol. Sep. 9 (1997) 63.
[140] S. T. Lee and S. V. Olesik, J. Chromatogr. A 707 (1995) 217.
[141] Y. Cui and S. V. Olesik, J. Chromatogr. A 691 (1995) 151.
[142] J. H. Knox, J. Chromatogr. Sci. 18 (1980) 453.
[143] J. H. Knox and M. T. Gilbert, J. Chromatogr. 186 (1979) 405.
[144] J. H. Knox, J. Chromatogr. Sci. 15 (1977) 352.
[145] P. A. Bristow and J. H. Knox, Chromatographia 10 (1977) 279.
[146] C. F. Poole and S. K. Poole, Anal. Chim. Acta 216 (1989) 109.
[147] C. R. Wilke and P. Chang, Am. Inst. Chem. Eng. J. 1 (1955) 264.
[148] R. T. Kennedy and J. W. Jorgenson, Anal. Chem. 61 (1989) 1128.
[149] W. Li, D. Pyo, Y. Wan, E. Ibanez, A. Malik and M. L. Lee, J. Microcol. Sep. 8 (1996) 259.
[150] J. P. C. Vissers, H. A. Claessens and C. A. Cramers, J. Chromatogr. A 779 (1997) 1.
[151] A. Berthod, J. Liq. Chromatogr. 12 (1989) 1169 and 1187.
[152] P. W. Carr and L. Sun, J. Microcol. Sep. 10 (1998) 149.
[153] B. A. Bidlingmeyer and F. V. Warren, Anal. Chem. 56 (1984) 1583A.
[154] H. Poppe, J. Chromatogr. 778 (1997) 3.
[155] R. Swart, J. C. Kraak and H. Poppe, Trends Anal. Chem. 16 (1997) 332.
[156] J. C. Sternberg, Adv. Chromatogr. 2 (1966) 205.
[157] F. W. Freebairn and J. H. Knox, Chromatographia 19 (1985) 37.
[158] H. H. Lauer and G. P. Rozing, Chromatographia 19 (1985) 641.
[159] M. V. Novotny and D. Ishii (Eds.), *Microcolumn Separations. Columns, Instrumentation and Ancillary Techniques*, Elsevier, Amsterdam, 1985.
[160] A. K. Bemgard and A. L. Colmsjo, J. High Resolut. Chromatogr. 13 (1990) 689.
[161] C. A. Lucy, L. L. M. Glavina and F. F. Cantwell, J. Chem. Edu. 72 (1995) 367.
[162] H. H. Lauer and G. P. Rozing, Chromatographia 14 (1981) 641.
[163] S. R. Bakalyar, C. Phipps, B. Spruce and K. Olsen, J. Chromatogr. A 762 (1997) 167.
[164] R. Tijssen, Sep. Sci. Technol. 13 (1978) 681.
[165] E. D. Katz and R. P. W. Scott, J. Chromatogr. 268 (1983) 169.
[166] J. G. Atwood and M. J. Golay, J. Chromatogr. 218 (1981) 97.
[167] K. A. Cohen and J. D. Stuart, J. Chromatogr. Sci. 25 (1987) 381.
[168] H. A. Claessens, C. A. Cramers and M. A. Kuyken, Chromatographia 23 (1987) 189.
[169] H. Poppe, J. Chromatogr. A 656 (1993) 19.

[170] F. Roubani-Kalantzopoulou, J. Chromatogr. A 806 (1998) 293.
[171] G. Gotmar, T. Fornstedt and G. Guiochon, J. Chromatogr. A 831 (1999) 17.
[172] F. James, M. Sepulveda, F. Charton, I. Quinones and G. Guiochon, Chem. Eng. Sci. 54 (1999) 1677.
[173] M. Jaroniec, J. Chromatogr. A 722 (1996) 19.
[174] N. Felitsyn and F. F. Cantwell, Anal. Chem. 71 (1999) 1862.
[175] P. Jandera, D. Komers, L. Andel and L. Prokes, J. Chromatogr. A 831 (1999) 131.
[176] K. Mihlbachler, K. Kaczmarski, A. Seidel-Morgenstern and G. Guiochon, J. Chromarogr. A 955 (2002) 35.
[177] C. Blumel, P. Hugo, A. Seidel-Morgenstern, J. Chromatogr. A 865 (1999) 51.
[178] O. Lisec, P. Hugo and A. Seidel-Morgenstern, J. Chromatogr. A 908 (2001) 19.
[179] P. Jandera, S. Buncekova, K. Mihlbachler, G. Guiochon, V. Backovska and J. Planeta, J. Chromatogr. A 925 (2001) 19.
[180] H. Guan and G. Guiochon, J. Chromatogr. A 724 (1996) 39.
[181] P. Jandera, M. Skavrada, L. Andel, D. Komers and G. Guiochon, J. Chromatogr. A 908 (2001) 3.
[182] L. Zhang, J. Selker, A. Qu and A. Velayudhan, J. Chromatogr. A 934 (2001) 13.
[183] A. Felinger, *Data Analysis and Signal Processing in Chromatography*, Elsevier, Amsterdam, 1998.
[184] N. Dyson, *Chromatographic Integration Methods*, The Royal Society of Chemistry, Cambridge, UK, 1998.
[185] V. B. Di Marco and G. G. Bombi, J. Chromatogr. A 931 (2001) 1.
[186] J. R. Conder, J. High Resolut. Chromatogr. 5 (1982) 341 and 397.
[187] D. J. Anderson and R. R. Walters, J. Chromatogr. Sci. 22 (1984) 353.
[188] W. W. Yau, S. W. Rementer, J. M. Boyajian, J. J. DeStefano, J. F. Graff, K. B. Lim and J. J. Kirkland, J. Chromatogr. 630 (1993) 69.
[189] M. S. Jeansonne and J. P. Foley, J. Chromatogr. Sci. 29 (1991) 258.
[190] J. O. Grimalt and J. Olive, Anal. Chim. Acta 248 (1991) 59.
[191] A. Berthod, Anal. Chem. 63 (1991) 1879.
[192] S. Le Vent, Anal. Chim. Acta 312 (1995) 263.
[193] J. W. Li, Anal. Chem. 69 (1997) 4452.
[194] K. Lan and J. W. Jorgenson, J. Chromatogr. 915 (2001) 1.
[195] J. Li, J. Chromatogr. A 952 (2002) 63.
[196] W. M. A. Niessen, H. P. M. Van Vliet and H. Poppe, Chromatographia 20 (1985) 357.
[197] W. P. N. Fernando and C. F. Poole, J. Planar Chromatogr. 4 (1991) 278.
[198] L. R. Snyder, J. J. Kirkland and J. L. Glajch, *Practical HPLC Method Development*, Wiley, New York, 1997.
[199] K. Suematsu and T. Okamoto, J. Chromatogr. Sci. 27 (1989) 13.
[200] P. Sandra, J. High Resolut. Chromatogr. 12 (1989) 82.
[201] J. P. Foley, Analyst 116 (1991) 1275.
[202] P. J. Schoenmakers, *Optimization of Chromatographic Selectivity. A Guide to Method Development*, Elsevier, Amsterdam, 1986.
[203] J. C. Berridge, Chemomet. Intel. Labor. Syst. 3 (1988) 175.
[204] A. Peeters, L. Buydens, D. L. Massart and P. J. Schoenmakers, Chromatographia 26 (1988) 101
[205] R. Cela, E. Leira, O. Cabaleiro and M. Lores, Comput. Chem. 20 (1996) 285.
[206] V. M. Morris, J. G. Hughes and P. J. Marriott, J. Chromatogr. A 755 (1996) 235.
[207] P. F. Vanbel, B. L. Tilquin and P. J. Schoenmakers, Chemomet. Intel. Labor. Systs. 35 (1996) 67.
[208] E. J. Klein and S. L. Rivera, J. Liq. Chromatogr. Rel. Technol. 23 (2000) 2097.
[209] A.-M. Siouffi and R. Phan-Tan-Luu, J. Chromatogr. A 892 (2000) 75.
[210] E. Grushka, J. Chromatogr. 316 (1984) 81.
[211] Y. Shen and M. L. Lee, Anal. Chem. 70 (1998) 3853.
[212] J. C. Medina, N. J. Wu and M. L. Lee, Anal. Chem. 73 (2001) 1301.
[213] J. Krupcik, T. Hevesi and P. Sandra, Collect. Czech. Chem. Commun. 60 (1995) 559.
[214] J. C. Giddings, *Unified Separation Science*, Wiley, New York, 1991.

[215] K. Lan and J. W. Jorgenson, Anal. Chem. 71 (1999) 709.
[216] M. Martin, Fresenius J. Anal. Chem. 352 (1995) 625.
[217] L. S. Ettre, Chromatographia 8 (1975) 291.
[218] K. Grob and G. Grob, J. Chromatogr. 207 (1981) 291.
[219] L. A. Jones, C. D. Burton, T. A. Dean, T. M. Gerig and J. R. Cook, Anal. Chem. 59 (1987) 1179.
[220] P. G. Muijselaar, M. A. van Stratten, H. A. Claessons and C. A. Cramers, J. Chromatogr. A 766 (1997) 187.
[221] S. Kolb, T. Welsch and J. P. Kutter, J. High Resolut. Chromatogr. 21 (1998) 435.
[222] L. M. Blumberg and M. S. Klee, J. Chromatogr. A 933 (2001) 1 and 13.
[223] H. Cortes (Ed.), *Multidimensional Chromatography: Techniques and Applications*, Marcel Dekker, New York, 1990.
[224] J. C. Giddings, J. Chromatogr. A 703 (1995) 3.
[225] Z. Liu, D. G. Patterson and M. L. Lee, Anal. Chem. 67 (1995) 3840.
[226] J. M. Davis, Adv. Chromatogr. 34 (1994) 109.
[227] A. Felinger and M. C. Pietrogrande, Anal. Chem. 73 (2001) 619A.
[228] J. M. Davis, M. Pompe and C. Samuel, Anal. Chem. 72 (2000) 5700.
[229] J. M. Davis, Anal. Chem. 69 (1997) 3796.
[230] J. M. Davis, J. Chromatogr. A 831 (1999) 37.
[231] C. Samuel and J. M. Davis, J. Chromatogr. A 842 (1999) 65.
[232] F. Dondi, A. Bassi, A. Cavazzini and M. C. Pietrogrande, Anal. Chem. 70 (1998) 766.
[233] F. Dondi, M. C. Pietrogrande and A. Felinger, Chromatographia 45 (1997) 435.
[234] A Felinger, Adv. Chromatogr. 39 (1998) 201.
[235] J. J. Kirkland, J. Chromatogr. Sci. 38 (2000) 535.
[236] R. T. Kennedy, I. German, J. E. Thompson and S. R. Witowski, Chem. Rev. 99 (1999) 3081.
[237] L. M. Blumberg, J. High Resolut. Chromatogr. 20 (1997) 679.
[238] L. M. Blumberg, J. High Resolut. Chromatogr. 22 (1999) 403.
[239] C. A. Cramers and P. A. Leclercq, J. Chromatogr. A 842 (1999) 3.
[240] Y. F. Shen and M. L. Lee, J. Microcol. Sep. 9 (1997) 21.
[241] M. van Lieshout, M. van Deursen, R. Derks, H. G. Janssen and C. Cramers, J. Microcol. Sep. 11 (1999) 155.
[242] M. van Deursen, H.-G. Janssen, J. Beens, P. Lipman, R. Reinierkens, G. Rutten and C. Cramers, J. Microcol. Sep. 12 (2000) 613.
[243] J. de Zeeuw, J. Peene, H.-G. Janssen and X. Lou, J. High Resolut. Chromatogr. 23 (2000) 677.
[244] T. M. Nahir and J. A. Gerbec, J. Chromatogr. A 915 (2001) 265.
[245] K. Mastovska, S. J. Lehotay and J. Hajslova, J. Chromatogr. A 926 (2001) 291.
[246] R. Annino, J. High Resolut. Chromatogr. 19 (1996) 285.
[247] R. Sacks, H. Smith and M. Nowak, Anal. Chem. 70 (1998) 29A.
[248] M. van Deursen, M. van Lieshout, R. Derks, H. G. Janssen and C. Cramers, J. High Resolut. Chromatogr. 22 (1999) 155.
[249] M. van Lieshout, R. Derks, H. G. Janssen and C. A. Cramers, J. High Resolut. Chromatogr. 21 (1998) 583
[250] F. David, D. R. Gere, F. Scanlan and P. Sandra, J. Chromatogr. A 842 (1999) 309.
[251] C. A. Monnig, D. M. Dohmeier and J. W. Jorgenson, Anal. Chem. 63 (1991) 807.
[252] M. C. Muller, M. Caude, J. E. Dauphin, L. Lecointre and J. Saint-Germain, Chromatographia 40 (1995) 394.
[253] D. J. Phillips, M. Capparella, M. Neue and Z. El Fallah, J. Pharm. Biomed. Anal. 15 (1997) 1389.
[254] I. M. Mutton, Chromatographia 47 (1998) 291.
[255] H. K. Lim, S. Stellingweif, S. Sisenwine and K. W. Chan, J. Chromatogr. A 831 (1999) 227.
[256] G. Mayr and T. Welsch, J. Chromatogr. A 845 (1999) 155.
[257] U. D. Neue and J. R. Mazzeo, J. Sep. Sci. 24 (2001) 921.
[258] K. Valko (Ed.), *Separation Methods in Drug Synthesis and Purification*, Elsevier, Amsterdam, 2000.

[259] L. R. Snyder, M. A. Stadalius and M. A. Quarry, Anal. Chem. 55 (1983) 1412A.
[260] Y. Shen and M. L. Lee, Chromatographia 45 (1997) 67.
[261] A. W. Moore and J. W. Jorgenson, Anal. Chem. 65 (1993) 3550.
[262] C. S. Effenhauser, A. Manz and H. M. Widmer, Anal. Chem. 65 (1993) 2637.
[263] X. Y. Sun, H. Singh, B. Millier, C. H. Warren and W. A. Aue, J. Chromatogr. A 687 (1994) 259.
[264] E. Katz (Ed.), *Quantitaive Analysis Using Chromatographic Techniques*, Wiley, Chichester, 1987.
[265] M. F. Delaney, Analyst 107 (1982) 606.
[266] D. Jin and H. L. Pardue, Anal. Chim. Acta 422 (2000) 1 and 11.
[267] A. N. Papas, CRC Crit. Revs. Anal. Chem. 20 (1989) 359.
[268] N. Dyson, J. Chromatogr. A 842 (1999) 321.
[269] J. C. Reijenga, J. Chromatogr. 585 (1991) 160.
[270] V. A. Meyer, Adv. Chromatogr. 35 (1995) 383.
[271] A. Felinger and G. Guiochon, J. Chromatogr. A 913 (2001) 221.
[272] A. N. Papas and T. P. Tougas, Anal. Chem. 62 (1990) 234.
[273] Y. Hayashi and R. Matsuda, Chromatographia 41 (1995) 66.
[274] M. L. Phillips and R. L. White, J. Chromatogr. Sci. 35 (1997) 75.
[275] B. Schirm and H. Watzig, Chromatographia 48 (1998) 331.
[276] J. Li, Anal. Chim. Acta 388 (1999) 187.
[277] S. Jurt, M. Schar and V. R. Meyer, J. Chromatogr. A 929 (2001) 165.
[278] V. R. Meyer, Chromatographia 40 (1998) 15.
[279] Y. Hayashi and R. Matsuda, Chromatographia 41 (1995) 75.
[280] V. R. Meyer, J. Chromatogr. Sci. 33 (1995) 26.
[281] V. J. Barwick, J. Chromatogr. A 849 (1999) 13.
[282] V. J. Barwick, S. L. R. Ellison, C. L. Lucking and M. J. Burn, J. Chromatogr. A 918 (2001) 267.
[283] D. B. Hibbert, J. X. Jiang and M. I. Mulholland, Anal. Chim. Acta 443 (2001) 205.
[284] M. A. Castillo and R. C. Castells, J. Chromatogr. A 921 (2001) 121.
[285] L. Cuadros-Rodriguez, A. Gonzalez-Casado, A. M. Garcia-Campana and J. L. Vilchez, Chromatographia 47 (1998) 550.
[286] P. Haefelfinger, J. Chromatogr. 218 (1981) 73.
[287] R. E. Pauls and R. W. McCoy, J. High Resolut. Chromatogr. 9 (1986) 600.
[288] Y. Hayashi and R. Matsuda, Anal. Sci. 11 (1995) 389.
[289] C. Nerin, J. Cacho, A. R. Tornes and I. Echarri, J. Chromatogr. A 672 (1994) 159.
[290] J. J. Sun and D. A. Roston, J. Chromatogr. A 673 (1994) 211.
[291] M. Bader, J. Chem. Edu. 57 (1980)703.

# Chapter 2

# The Column in Gas Chromatography

## 2.1 INTRODUCTION

Separations are possible in gas chromatography if the solutes differ in their vapor pressure and/or intensity of solute-stationary phase interactions. As a minimum requirement the sample, or some convenient derivative of it, must be thermally stable at the temperature required for vaporization. The fundamental limit for sample suitability is established by the thermal stability of the sample and system suitability by the thermal stability of column materials. In contemporary practice an upper temperature limit of about 400°C and a sample molecular weight less than 1000 is indicated, although higher temperatures have been used and higher molecular weight samples have been separated in a few instances.

When a solid adsorbent serves as the stationary phase the separation technique is called gas-solid chromatography (GSC). When the stationary phase is a liquid spread on an inert support or coated as a thin film onto the wall of a capillary column the separation technique is called gas-liquid chromatography (GLC). In packed column gas chromatography the separation medium is a coarse powder (coated with a liquid for GLC) through which the carrier gas flows. If the adsorbent, liquid phase, or both, are coated onto the wall of a narrow bore column of capillary dimensions, the technique is called wall-coated open tubular (WCOT), support-coated open tubular (SCOT), or porous-layer open tubular (PLOT) column gas chromatography. As indicated by name the characteristic feature of open-tubular columns is their high permeability created by the open passageway through the center of the column.

Five column types have been used in gas chromatography. Classical packed columns have internal diameters greater than 2 mm and are packed with particles in the range 100 to 250 μm. Micropacked columns have diameters less than 1 mm and a similar packing density to classical packed columns ($d_P$ / $d_C$ less than 0.3, where $d_P$ is the average particle diameter and $d_C$ the column diameter). Packed capillary columns have a column diameter less than 0.6 mm and are packed with particles of 5-20 μm diameter. SCOT

Table 2.1
Representative properties of different column types for gas chromatography
$H_{min}$ = minimum plate height at the optimum mobile phase velocity $u_{opt}$

| Column type | Phase ratio | $H_{min}$ (mm) | $u_{opt}$ (cm/s) | Permeability ($10^7$.cm$^2$) |
|---|---|---|---|---|
| Classical Packed | 4-200 | 0.5-2 | 5-15 | 1-50 |
| Micropacked | 50-200 | 0.02-1 | 5-10 | 1-100 |
| Packed Capillary | 10-300 | 0.05-2 | 5-25 | 5-50 |
| SCOT | 20-300 | 0.5-1 | 10-100 | 200-1000 |
| WCOT | 15-500 | 0.03-0.8 | 10-100 | 300-20000 |

columns are capillary columns containing a liquid phase coated on a surface covered with a layer of solid support material, leaving an open passageway through the center of the column. (A PLOT column is simply a SCOT column without the liquid phase). In WCOT columns the liquid phase is coated directly on the smooth or chemically etched column wall. Some characteristic properties of the different column types are summarized in Table 2.1 [1]. The most significant difference among the various column types is their permeability. The open tubular columns offer much lower flow resistance and can be used in much longer lengths to obtain very high total plate numbers. The minimum plate height of the best packed column in gas chromatography is about 2-3 particle diameters whereas that of an open tubular column will be similar to the column diameter. The intrinsic efficiency of open tubular columns, therefore, is not necessarily greater than that of a packed column, but because of their greater permeability at a fixed column pressure drop a greater total plate number can be obtained, since longer columns can be used.

The phase ratio (ratio of the volume of gas to liquid phase) for a number of typical column configurations is given in Table 2.2 [2]. At a constant temperature the partition coefficient will be the same for all columns prepared from the same stationary phase. Consequently, for a column with a large phase ratio the retention factor will be small and *vice versa*. The plate number required for a separation becomes very large at small retention factor values (see Table 1.10). Columns with low phase ratios, that is thick-film columns, have a lower intrinsic efficiency than thin-film columns but provide better resolution of low boiling compounds, because they provide more favorable retention factors. They also allow the separation to be performed in a higher and more convenient temperature range than is possible with thin-film columns. The opposite arguments apply to high boiling compounds that have long separation times on thick-film columns because their retention factors are too large. Increasing the phase ratio lowers the retention factors to a value within the optimum range so that there is little deterioration in resolution and a substantial saving in separation time. Packed columns have low phase ratios compared to WCOT columns leading to longer separation times at a constant temperature. SCOT columns occupy an intermediate position. Since several combinations of film thickness and column radius can be used to generate the same phase ratio, there are other factors that need to be considered for selecting these variables for a particular separation.

Table 2.2
Characteristic properties of some representative columns
$H_{min}$ is the minimum plate height and k is for undecane at 130°C

| Column type | Length (m) | Internal diameter (mm) | Film thickness (μm) | Phase ratio | Retention factor (k) | $H_{min}$ (mm) | Column plate number | Plates per meter |
|---|---|---|---|---|---|---|---|---|
| Classical | 2 | 2.16 | 10%(w/w) | 12 | 10.4 | 0.55 | 3,640 | 1,820 |
| packed | 2 | 2.16 | 5%(w/w) | 26 | 4.8 | 0.50 | 4,000 | 2,000 |
| SCOT | 15 | 0.50 | | 20 | 6.2 | 0.95 | 15,790 | 1,050 |
| | 15 | 0.50 | | 65 | 1.9 | 0.55 | 27,270 | 1,820 |
| WCOT | 30 | 0.10 | 0.10 | 249 | 0.5 | 0.06 | 480,000 | 16,000 |
| | 30 | 0.10 | 0.25 | 99 | 1.3 | 0.08 | 368,550 | 12,285 |
| | 30 | 0.25 | 0.25 | 249 | 0.5 | 0.16 | 192,000 | 6,400 |
| | 30 | 0.32 | 0.32 | 249 | 0.5 | 0.20 | 150,000 | 5,000 |
| | 30 | 0.32 | 0.50 | 159 | 0.8 | 0.23 | 131,330 | 4,380 |
| | 30 | 0.32 | 1.00 | 79 | 1.6 | 0.29 | 102,080 | 3,400 |
| | 30 | 0.32 | 5.00 | 15 | 8.3 | 0.44 | 68,970 | 2,300 |
| | 30 | 0.53 | 1.00 | 132 | 0.9 | 0.43 | 70,420 | 2,340 |
| | 30 | 0.53 | 5.00 | 26 | 4.8 | 0.68 | 43,940 | 1,470 |

In general, increasing the column radius and film thickness for a WCOT column will lead to an increase in the column plate height and decrease in efficiency. However, the relationship between the variables is complex (see section 1.5.2) and depends on the retention factor since this term appears explicitly in the contribution from resistance to mass transfer to the plate height. For thin-film columns ($d_f < 0.25$ μm) resistance to mass transfer in the stationary phase is small (frequently negligible) so that decreasing the radius of thin-film columns by minimizing the mobile phase mass transfer term leads to increased efficiency. Narrow-bore, thin-film columns are the most intrinsically efficient provided that the inlet pressure is not limited. If the column radius is held constant and the film thickness is increased the column efficiency will decline because resistance to mass transfer in the stationary phase will eventually dominate the plate height equation.

Today, WCOT columns and classical packed columns dominate the practice of gas-liquid chromatography and PLOT columns and classical packed columns the practice of gas-solid chromatography. WCOT and PLOT columns are the first choice for analytical separations because of their superior peak capacity and, in the case of WCOT columns, greater chemical inertness. Packed columns are less expensive, require little training in their use, are better suited to isolating preparative-scale quantities, and can better tolerate samples containing involatile or thermally labile components. Only a limited number of stationary phases have been immobilized successfully in open tubular columns compared to the large number of phases available for use in packed columns. Packed columns are still used in many fundamental studies of solvent properties because of the ease of column preparation and the large amounts of solvent that can be accommodated to minimize measurement errors. The attendant difficulties of high-pressure operation and in uniformly packing long columns have conspired to make micropacked columns the least popular choice for analytical applications [3,4]. Micropacked columns have been used in monitoring extraterrestrial atmospheres because of their favorable combination of durability, high sample capacity and low mobile phase flow rates [5,6]. SCOT columns have more or less been totally replaced by thick-film WCOT columns. They might still prove useful for applications that demand a large sample capacity or use of selective stationary phases that are difficult to coat on open tubular columns or for separations that require a greater plate number than can be provided by classical packed columns. Slurry packed capillary columns with small diameter particles (5-20 μm) have been used for fast separations in instruments modified for high-pressure operation [7-9]. These columns provide higher efficiency per unit time, greater retention and a higher sample capacity than WCOT columns but their low permeability limits the useful column length. Consequently, the total plate number that can be achieved is small (< 20,000). They may prove useful for fast separations of simple mixtures but this has to be balanced against the inconvenience of high-pressure operation.

## 2.2 MOBILE PHASES

It is convenient to divide mobile phases for gas chromatography into solvating and non-solvating gases. The latter are the most important and dominate the practice of gas chromatography. Common carrier gases are hydrogen, helium and nitrogen. These gases behave almost ideally at the low pressures and typical temperatures used in gas chromatography. Consequently, they do not influence selectivity in gas-liquid chromatography except for minor contributions attributed to nonideal gas behavior at extreme operating conditions. In gas-solid chromatography competition between solute molecules and carrier gas molecules for sorbent surface active sites can result in significant changes in retention, particularly for heavier gases like carbon dioxide [10,11]. In this case, retention factors also depend on the average gas pressure for the column. In gas-liquid chromatography the primary function of the carrier gas is to provide the transport mechanism for the sample through the column. The choice of carrier gas may still influence resolution, however, through its effect on column efficiency, which arises from differences in solute diffusion rates. It can also affect separation time, because the optimum carrier gas velocity decreases with the decrease in solute diffusivity. In pressure-limiting conditions, gas viscosity differences are important as well. Other considerations for choosing a particular carrier gas might be cost, purity, safety, reactivity and detector compatibility. These factors also tend to favor the inert gases, nitrogen, hydrogen and carbon dioxide.

The viscosity of the carrier gas determines the column pressure drop for a given velocity. For hydrogen, helium, and nitrogen the viscosity can be calculated for conditions germane to gas chromatography by Eq. (2.1) [12,13].

$$\eta_t = \eta_o(T_t / T_o)^n \tag{2.1}$$

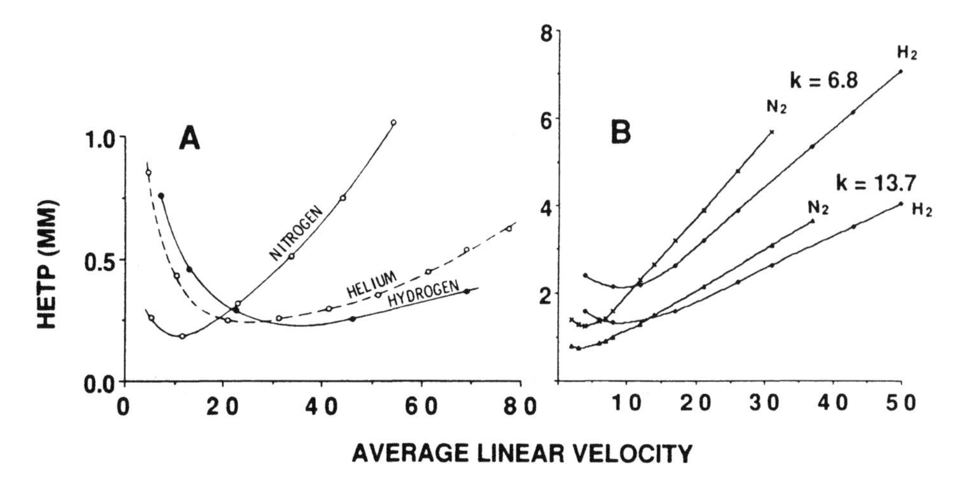


Figure 2.1. Van Deemter plots indicating the influence of the choice of carrier gas on column efficiency for thin-film (A) and thick-film (B) open tubular columns for solutes with different retention factors.

where $\eta_t$ is the gas viscosity in μP.s (=μN.s/m$^2$) at temperature t, $\eta_o$ the gas viscosity at 0°C, and T (K) the temperature. The reference values for the three gases are hydrogen ($\eta_o$ = 8.3687 and n = 0.6961), helium ($\eta_o$ = 18.6804 and n = 0.6977) and nitrogen ($\eta_o$ = 16.821 and n = 0.7167). Tabulated values of gas viscosity at 2°C intervals for the temperature range 0 to 400°C are given in [13]. Differences between the viscosity calculated by Eq. (2.1) and the more exact (although complex) equations derived from the kinetic theory of gases are < 1% for the temperature range indicated. For the three gases, corrections for pressure differences are negligible up to an inlet pressure of at least 5 atm. Viscosity increases with temperature for gases, causing a decrease in the carrier gas velocity at constant pressure. The rate of change of viscosity with temperature is approximately the same for the three gases. Fused silica, commonly used in the preparation of open tubular columns, is significantly more permeable to helium than either hydrogen or nitrogen [14]. For most circumstances this is unimportant, but for long columns at high temperatures significant loss of helium through the column wall is observed. Column hold-up times can be up to 15% less than predicted by theory but typical deviations are usually only a few percent.

Solute diffusivity influences both the plate height and the position of the optimum carrier gas velocity described by the van Deemter equation [15-18]. Nitrogen provides the lowest plate height, Figure 2.1, but it occurs at an optimum velocity, which is rather low leading to long separation times. In the optimum plate height region the ascending portions of the curve are much shallower for hydrogen and helium. Thus for separations at mobile phase velocities higher than the optimum velocity, compromise conditions selected to minimize separation times, hydrogen, and to a lesser extent helium, show only a modest sacrifice in column efficiency compared to nitrogen.

The above observations are sound for thin-film open tubular columns ($d_f$ < 0.5 μm) and low loaded packed columns. Hydrogen is clearly the preferred carrier gas when efficiency and separation time are considered together. For thick-film columns different conclusions are reached, since diffusion in the stationary phase contributes significantly to the band broadening mechanism. In this case the retention factor, temperature, and stationary phase diffusion coefficient have to be considered when selecting a carrier gas for a particular separation [19]. In going from a thin-film to a thick-film column (which corresponds to reducing the phase ratio) the efficiency of the columns, as measured by the minimum plate height, decreases substantially and depends on the carrier gas identity. The optimum carrier gas velocity is reduced by perhaps a factor of two and also depends on the carrier gas identity. The slopes of the ascending portion of the van Deemter curves at high velocity are steeper for the thick-film columns and depend primarily on the stationary phase diffusion coefficient (Figure 2.1). For maximum efficiency thick-film columns should be operated at their optimum mobile phase velocity with nitrogen as the carrier gas. These conditions will enhance resolution while compromising separation time.

For packed columns, resistance to mass transfer in the stationary phase is always significant at normal phase loadings. For pressure limited conditions hydrogen is preferred, because its viscosity is only about half that of helium and nitrogen. Nitrogen provides (slightly) higher efficiency at low temperatures and low flow rates, while hydrogen is superior at higher temperatures and at above optimum flow rates [15,20].

The only disadvantage to the use of hydrogen as a carrier gas is the real or perceived explosion hazard from leaks within the column oven. Experience has shown that the conditions required for a catastrophic explosion may never be achieved in practice with forced air convection ovens. However, commercially available gas sensors will automatically switch off the column oven and carrier gas flow at air-hydrogen mixtures well below the explosion threshold limit. A considerable difference in the relative cost of helium in the USA and Europe has resulted in different preferences on the two continents. For open tubular columns helium is widely used in the USA for safety rather than theoretical considerations while hydrogen is commonly used in Europe.

### 2.2.1 Solvating Mobile Phases

Occasionally organic vapors or other gases are added to one of the common carrier gases to modify the column separation characteristics [21-25]. Typical examples are steam, ammonia, nitrous oxide, carbon dioxide and formic acid. In general, these gases and their mixtures do not behave ideally and they can influence separations through the solvating capacity of the mobile phase. This is unlikely their main mode of action, however, and changes in peak shape and relative retention caused by masking the adsorption of polar solutes and formation of binary liquid phases through their solubility in the stationary phase are probably more important. For these system selectivity can often be adjusted by changing the partial pressure of the organic vapor in the carrier gas and by varying the column average pressure. In general, however, the inconvenient experimental arrangement, detector incompatibility and limited temperature range for effective application, combined with the possibility of achieving the desired separations by other conventional means have limited interest in this approach. A more radical approach is represented by solvating gas chromatography, which employs a transition from supercritical fluid to gas phase conditions along the column to modify selectivity (section 7.7.3).

## 2.3 STATIONARY PHASES

Given the non-solvating properties of the carrier gases commonly used in gas-liquid chromatography optimization of selectivity requires selection of an appropriate stationary phase. Over the history of gas-liquid chromatography thousands of substances have been used as stationary phases [26-30]. For several reasons most of these have been abandoned. Many liquids were technical products of variable or undefined composition unable to meet increasing demands for retention reproducibility. Improved methods of classification indicated that many liquids had similar separation properties and after grouping into classes a few liquids characteristic of each class could suffice for method development. Other liquids had poor thermal stability or high vapor pressure and were replaced by materials with more suitable physical and similar chromatographic properties. Of great significance was the gradual evolution of column technology resulting in increased use of open tubular columns and the preferential selection of liquids that could form stable films on the walls of these columns. This required rather specific properties that many liquids suitable for packed column gas chromatography could not provide. It would not serve our purpose to comprehensively review all available material since much could be arguably identified as redundant. If only the most recent advances were described, though, the reader would be left with no basis to interpret early studies in the field that are still important in piloting the development of column technology. We will attempt to find a middle ground reflecting current practice and a flavor of the recent past.

General considerations that influence the choice of a particular liquid for use as a stationary phase in gas-liquid chromatography include the following. Practical considerations dictate that the liquid should be unreactive, of low vapor pressure, have good coating characteristics (wet the materials used in column fabrication) and have reasonable solubility in some common volatile organic solvent. It is desirable that the liquid phase has a wide temperature operating range covering as far as possible the full temperature range (-60° to 400°C) used in gas-liquid chromatography. The lower operating temperature for a liquid phase is usually its melting point, or glass transition temperature if a polymer. The lowest temperature at which the mass transfer properties of the melted solid provide adequate column efficiency is an alternative specification for the lowest useful operating temperature. The maximum allowable operating temperature is usually determined by the thermal stability of the stationary phase or its vapor pressure. In practice, the highest temperature that can be maintained

Table 2.3
Characteristic properties of some liquid phases used in packed column gas chromatography
PPE-5 = Poly(phenyl ether), EGS = poly(ethylene glycol succinate) and DEGS = poly(diethylene glycol succinate), Carbowax 20M = poly(ethylene glycol) and FFAP = Carbowax 20M treated with 2-nitroterephthalic acid

| Name and structure | | Temperature range (°C) | |
|---|---|---|---|
| | | Minimum | Maximum |
| Hexadecane | $C_{16}H_{34}$ | < 20 | 50 |
| Squalane | 2,6,10,15,19,23-hexamethyltetracosane | < 20 | 120 |
| Apolane-87 | $(C_{18}H_{37})_2CH(CH_2)_4C(C_2H_5)_2(CH_2)_4CH(C_{18}H_{37})_2$ | 30 | 280 |
| Fomblin YR | $-[OCF(CF_3)CF_2)_n(OCF_2)_m]-$ | 30 | < 255 |
| PPE-5 | $C_6H_5O(C_6H_5O)_3C_6H_5$ | 20 | 200 |
| Dioctyl Phthalate | $C_6H_4(COOC_8H_{17})_2$ | < 20 | 160 |
| EGS | $HO(CH_2)_2[OOCCH_2CH_2COO(CH_2)_2]_nOH$ | 100 | 210 |
| DEGS | $HO(CH_2)_2O(CH_2)_2[OOCCH_2CH_2COO(CH_2)_2O(CH_2)_2]_nOH$ | 20 | 200 |
| Carbowax 20M | $HO(CH_2CH_2O)_nCH_2CH_2OH$ | 60 | 225 |
| FFAP | | 50 | 250 |
| 1,2,3-Tris(2-cyanoethoxy)propane | $(CH_2OCH_2CH_2CN)_3$ | 20 | 170 |
| Tetrabutylammonium Perfluorooctanesulfonate | | < 20 | 220 |
| Tetrabutylammonium 4-Toluenesulfonate | | 55 | 200 |
| Tetrabutylammonium Tetrafluoroborate | | 162 | 290 |
| Ethylammonium 4-Toluenesulfonate | | 121 | 220 |
| Tetrabutylphosphonium chloride | | 83 | 230 |

without breakup of the liquid film into droplets may be used. The liquid temperature range requirement generally favors the selection of polymeric materials, although the chemical structure and composition of polymeric phases is likely to show greater variability than a synthetic compound with a defined chemical structure.

### 2.3.1 Hydrocarbon and Perfluorocarbon Stationary Phases

High molecular weight hydrocarbons such as hexadecane, squalane ($C_{30}H_{62}$), Apolane-87 ($C_{87}H_{176}$) and Apiezon greases have long been used as nonpolar stationary phases in gas chromatography, Table 2.3. Squalane is obtained by the complete hydrogenation of squalene isolated from shark liver oil. As a natural product its chemical purity is sometimes doubtful with squalene and batyl alcohol being the principal impurities [31]. The Apiezon greases are prepared by the high temperature treatment and molecular distillation of lubricating oils. They are of ill-defined composition and contain unsaturated hydrocarbon groups as well as residual carbonyl and carboxylic acid groups [32]. Both squalane and Apiezon should be purified by chromatography over charcoal and alumina before use, and always if colored [31,33]. Exhaustive hydrogenation can be used to saturate aromatic and olefinic groups to produce a more stable product, Apiezon MH, having properties closer to those of a typical saturated hydrocarbon phase. Apolane-87 is a synthetic hydrocarbon and the most reproducible and thermally stable of the hydrocarbon phases [34]. All hydrocarbon phases are susceptible to oxidation by reaction with the small quantities of oxygen present in the carrier gas, particularly at elevated temperatures [31,35]. Oxidation alters the chromatographic properties of these phases by introducing polar, oxygenated functional groups and in extreme cases may cause scission of the carbon backbone resulting in a high level of column bleed. Apolane-87 is the most resistant of the common hydrocarbon phases to oxidation because of its low concentration of tertiary hydrogen atoms.

Hydrocarbon phases provide separations based on dispersion interactions with a variable contribution from induction interactions possible when separating polar solutes. For many years they were widely used in the petrochemical industry for speciating volatile hydrocarbon mixtures but have subsequently been replaced by specially synthesized poly(alkylsiloxane) phases containing large alkyl groups [36-38]. Hydrocarbon phases are used as nonpolar reference phases in various schemes proposed to measure the selectivity of polar phases. The method of retention index differences of Rohrschneider and McReynolds (section 2.4.4.2) and the free energy nonpolar interaction term of Kollie and Poole [39] used squalane as a reference phase and the log $L^{16}$ solute descriptor of the solvation parameter model (section 1.4.3) [40-43] is calculated from correlation equations based on hydrocarbon phases. Hydrocarbon phases have weak-support deactivating properties resulting in unsymmetrical peak shapes and sample-size dependent retention on undeactivated packed column supports [31,42,44]. Even on silanized supports wetting problems persist and a definite transition at a minimum phase loading corresponding to coalescence of the liquid droplets to a continuous film may be observed [31,45]. WCOT columns coated with hydrocarbon phases demonstrate poor temperature stability and limited lifetimes.

Sporadic accounts of the use of fluorocarbon liquid phases for the separation of reactive compounds or speciation of perfluorocarbons and Freons have appeared almost from the inception of gas chromatography [46-48]. Reactive chemicals such as metal fluorides, halogens, interhalogen compounds, and the halide compounds of hydrogen, sulfur, and phosphorus tend to destroy conventional phases. The strong carbon-fluorine bonds in highly fluorinated phases provide the necessary chemical resistance for these applications. Highly-fluorinated phases also posses greater selectivity for the separation of isomeric perfluorocarbons and Freons. The low cohesive energy of perfluorocarbon stationary phases compared with the corresponding hydrocarbon phases results in significant deviations from Raoult's law for fluorocarbon/hydrocarbon mixtures. Consequently, for highly fluorinated stationary phases retention will be less than that for conventional phases, allowing an extension of the molecular weight separating range of gas chromatography or the separation of thermally labile substances at lower temperatures [46,49]. The same weak intermolecular interactions are responsible for the poor support/wall wetting characteristics, low column efficiencies and low maximum operating temperature limits characteristic of early highly fluorinated stationary phases. These deficiencies can be overcome by modifying the structure of the phase to contain anchor groups capable of strong support/wall interactions. Typical of these modified phases are the poly(perfluoroalkyl ether) Fomblin oils [50,51], the fluorinated alkyl esters [52] and poly(diphenyl/1H,1H,2H,2H-perfluorodecylmethyl)siloxane phases [49]. The latter phase can be immobilized and bonded to the wall of glass capillary columns allowing temperatures up to 400°C to be used, for example, for the separation of beeswax.

### 2.3.2 Ether and Ester Stationary Phases

The dialkyl phthalates and tetrachlorophthalates are moderately polar and weakly hydrogen-bond basic liquid phases [53]. Phthalate esters with long alkyl chains are the less volatile but have lower selectivity for polar interactions. Phthalate esters with octyl to dodecyl alkyl chains are a reasonable compromise between the desire for a wide temperature operating range and useful solvent selectivity, Table 2.3. Still, their high volatility compared to other phases and limited selectivity for polar interactions have reduced their general importance. Earlier claims that tetrachlorophthalates form strong charge-transfer complexes with electron-donor compounds are probably not well founded.

The term polyester is used to describe a wide range of resinous composites derived from the reaction of a polybasic acid with a polyhydric alcohol [54]. The most frequently used polyester phases for gas chromatography are the succinate and adipate esters of ethylene glycol, diethylene glycol and butanediol, Table 2.3. These phases are used almost entirely in packed column gas chromatography for the separation of polar compounds. Changes in polymer composition by column conditioning, low tolerance to oxygen and water (particularly above 150°C), and solute exchange reactions involving the polyester functional groups and alcohols, acids, amines and esters have contributed to their diminished use in recent years. They have been replaced in many of their former applications by the more stable poly(cyanoalkylsiloxane) phases.

The meta-linked poly(phenyl ethers) are moderately polar liquid phases with a defined chemical structure. Their low volatility is exceptional for low molecular weight liquids. The five and six ring poly(phenyl ethers) have useful operating temperature ranges up to 200°C and 250°C, respectively. Poly(phenyl ethers) containing polar functional groups are less useful liquid phases due to poor column efficiencies and low thermal stability [55]. Copolymerization of the five or six ring poly(phenyl ethers) with diphenyl ether-4,4'-disulfonyl chloride yields a polyether sulfone of undefined molecular weight and chemical structure suitable for the separation of a wide range of high-boiling polar compounds at temperatures in the range 200-400°C [56].

Poly(ethylene glycols) have been widely used for the separation of volatile polar compounds, such as flavor and fragrance compounds, and fatty acid esters [57]. Carbowax 20M is the most popular phase for packed column separations, Table 2.3. It is a waxy solid with a molecular weight of 14,000 to 18,000 melting at about 60°C to a stable liquid with a maximum operating temperature of about 225°C [58]. Specially purified poly(ethylene glycols) of higher molecular weight such as Superox-4 (molecular weight 4 million) [59] and Superox-20M [59,60] can be used at temperatures up to about 250-275°C. Pluronic phases have a lower polydispersity than Carbowax phases and are prepared by condensing propylene oxide with propylene glycol [61]. The resulting chain is then extended on both sides by the addition of controlled amounts of ethylene oxide until the desired molecular weight is obtained. Pluronic liquid phases with average molecular weights of 2000 to 8000 are most useful for gas chromatography and can be used at temperatures up to about 220-260°C. Condensing Carbowax 20M with 2-nitroterephthalic acid produces a new phase, FFAP, that has been recommended for the separation of organic acids. This phase is unsuitable for the separation of basic compounds or aldehydes with which it reacts. The poly(ethylene glycols) are rapidly degraded by oxygen and moisture at high temperatures. Strong acids and Lewis acid catalysts may also degrade these polymers.

Carbowax 20M was one of the earliest phases used routinely in WCOT columns because of its good coating characteristics. The thermal decomposition products of Carbowax 20M provided a popular method of surface deactivation for subsequent coating with Carbowax 20M and other phases in the early development of capillary column technology [62]. The mechanism of Carbowax deactivation remains unknown, although it is believed that the degradation products bond chemically to the surface silanol groups at the high temperatures employed for deactivation. The method is not widely used today. The improved stability of modern poly(ethylene glycol) WCOT columns towards water and oxygen and the availability of higher operating temperatures is achieved through bonding the polymer to the column wall and/or through crosslinking. For commercial columns the details remain propriety information but some general clues are contained in the literature [61,63-65]. Poly(ethylene glycols) are more difficult to immobilize by free radical crosslinking than the poly(siloxane) phases. Immobilization is possible using "autocrosslinkable" poly(ethylene glycols), free radical crosslinking of copolymers of poly(ethylene glycol) and vinylsiloxanes or by acid catalyzed reaction with an alkyltrialkoxysilane coupling agents. The columns are coated with the poly(ethylene glycol) and necessary reagents and the immobilization achieved by simply heating the column in an inert atmosphere to between 150-200°C for the required time. Crosslinking partially destroys the crystallinity of the poly(ethylene glycol) polymer resulting in a lower column operating temperature and improved film diffusion properties. Diffusion in immobilized poly(ethylene glycol) phases is still less than that observed for immobilized poly(siloxane) phases and thinner films are often selected in practice to maintain acceptable separation efficiency.

### 2.3.3 Liquid Organic Salts

The liquid organic salts are a novel class of ionic liquids with good solvent properties for organic compounds [66-69]. The alkylammonium and alkylphosphonium salts with weak nucleophilic anions form thermally stable liquids with a low vapor pressure at temperatures exceeding their melting points by 150°C or more in favorable cases, Table 2.3. In addition, a number of these salts have low melting points including several that are liquid at room temperature. For the low molecular weight tetraalkylammonium salts the lowest melting points are often found for the tetrabutylammonium salts. Weak nucleophilic anions such as the sulfonate and tetrafluoroborate anions are generally the most stable and provide the widest liquid operating temperature range.

The liquid organic salts have favorable mass transfer properties and the efficiency of column packings prepared with liquid organic salts are not obviously different to those prepared from conventional non-ionic phases. The liquid organic salts possess low chemical reactivity and transformation reactions are relatively rare [66,70]. Nucleophilic displacement can occur for alkyl halides, degradation of alkanethiols was observed for some salts, and proton transfer and other acid/base reactions can affect the recovery of amines. The unique selectivity of the liquid organic salts is a result of their strong hydrogen-bond basicity and significant capacity for dipole-type interactions. In this respect their solvent properties cannot be duplicated by common non-ionic liquid phases. The retention properties of the salts can be correlated with the structure of the ions, particularly the extent of charge delocalization and intramolecular association, providing the possibility of designing salts for specific applications [68]. Liquid organic salts do not wet glass surfaces well but can be coated on whisker walled and sodium chloride modified glass surfaces for the preparation of WCOT columns. The low viscosity of the liquid organic salts restricts the range of available film thicknesses and the accessible temperature operating range of WCOT columns [66]. No mechanism has been indicated for the preparation of immobilized liquid organic salt columns.

### 2.3.4 Poly(siloxane) Stationary Phases

The poly(siloxanes) are the most popular liquid phases for gas chromatography. A concert of favorable properties that includes a wide temperature operating range (low temperature glass transition point, low vapor pressure and high thermal stability), acceptable diffusion and solubility properties for different solute types, chemical inertness, good film forming properties, ease of synthesis with a wide range of chromatographic selectivity, and ease of immobilization for applications employing open tubular columns largely accounts for their dominant position in column technology [30,71,72]. The poly(siloxanes) used in gas chromatography are generally linear polymers with structures given in Figure 2.2, in which R can be a number of different substituents such as methyl, vinyl, phenyl, 3,3,3-trifluoropropyl, or 3-cyanopropyl. By varying the identity and relative amount of each R group as well as the ratio of the monomer units n to m, polymers with a wide range of solvent properties can be prepared.

The poly(siloxane) polymers are usually prepared by the acid or base hydrolysis of appropriately substituted dichlorosilanes or dialkoxysilanes, or by the catalytic polymerization of small ring cyclic siloxanes [71-75]. The silanol-terminated polymers are suitable for use after fractionation or are thermally treated to increase molecular weight and in some cases endcapped by trimethylsilyl, alkoxy or acetyl groups [76,77]. Poly(siloxanes) synthesized in this way are limited to polymers that contain substituent groups that are able to survive the relatively harsh hydrolysis conditions, such as alkyl, phenyl, 3,3,3-trifluoropropyl groups. Hydrosilylation provides an alternative route to the synthesis of poly(siloxanes) with labile or complicated substituents (e.g. cyclodextrin, oligoethylene oxide, liquid crystal, amino acid ester, and alcohol) [78-81]. In this case

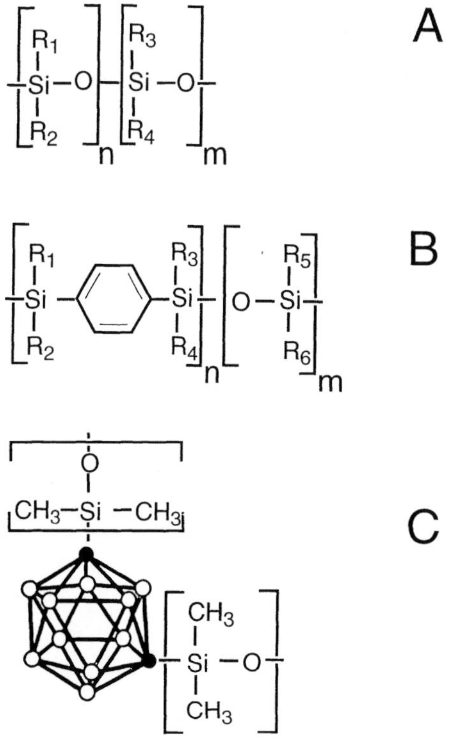

Figure 2.2. General structures of poly(siloxane) liquid phases. A, poly(siloxane) polymer; B, poly(silarylene-siloxane) copolymer; and C, a poly(carborane-siloxane) copolymer (● = carbon and ○ = BH).

synthesis starts with a poly(alkylhydrosiloxane) of the desired molecular weight in which the required fraction of R groups on the poly(siloxane) backbone are hydrogen. The substituent group to be incorporated into the polymer with a sidearm alkene group is then attached to the polymer at each Si-H bond by catalytic (chloroplatinic acid) hydrosilylation under relatively mild conditions. Condensation of carbofunctionalized dihalosilanes and oligodimethylsiloxanolate salts in aprotic solvents provides another route for the preparation of poly(siloxanes) with labile substituents [82]. The reactive carbofunctional groups (e.g. isocyanate, epoxy, activated carbonyl esters, etc.) provide an attachment point for the desired substituents by means of secondary reactions through formation of ester, amide, carbamate bonds, etc. Materials of high purity, particularly with regard to the presence of residual catalyst, and of narrow molecular weight range are required for chromatographic purposes. Although many methods are used for characterizing poly(siloxanes), multinuclear NMR is the most informative, providing information about the chemical composition, end group type, average molecular weight, branching, and the backbone microstructure [83].

Table 2.4
Characteristic properties of some poly(siloxane) liquid phases used for packed column gas chromatography

| Name | Structure | Viscosity (cP) | Average molecular weight | Temperature operating range (°C) Minimum | Maximum |
|---|---|---|---|---|---|
| OV-1 | Dimethylsiloxane | gum | > $10^6$ | 100 | 350 |
| OV-101 | Dimethylsiloxane | 1500 | 30,000 | <20 | 350 |
| OV-7 | Phenylmethyldimethylsiloxane<br>80 % methyl and 20 % phenyl | 500 | 10,000 | <20 | 350 |
| OV-17 | Phenylmethylsiloxane<br>50 % methyl and 50 % phenyl | 1300 | 40,000 | <20 | 350 |
| OV-25 | Phenylmethyldiphenylsiloxane<br>25 % methyl and 75 % phenyl | >100,000 | 10,000 | <20 | 300 |
| OV-210 | Trifluoropropylmethylsiloxane<br>50 % methyl and 50 % 3,3,3-trifluoropropyl | 10,000 | 200,000 | <20 | 275 |
| OV-225 | Cyanopropylmethylphenylmethylsiloxane<br>50 % methyl, 25 % phenyl and 25 % 3-cyanopropyl | 9000 | 8,000 | <20 | 250 |
| Silar 7CP | Cyanopropylphenylsiloxane<br>75 % 3-cyanopropyl and 25 % phenyl | | | 50 | 250 |
| OV-275 | Di(cyanoalkyl)siloxane<br>70 % 3-cyanopropyl and 30 % 2-cyanoethyl | 20,000 | 5,000 | | 250 |
| Silar 10CP | Di(3-Cyanopropyl)siloxane | | | 50 | 250 |

The gradual evolution of column technology revealed that the low to medium viscosity poly(siloxane) oils preferred as packed column stationary phases, Table 2.4, were unsuitable for preparing thermally stable and efficient open tubular columns. Gum phases of generally higher molecular weight, but more importantly higher viscosity, were required for the preparation of stable films resistant to disruption at elevated temperatures. Poly(dimethylsiloxanes) have a coiled helical structure with their methyl groups pointing outwards. Increasing temperature causes the mean intermolecular distance to increase but at the same time expansion of the helices occurs, tending to diminish this distance. As a result, the viscosity appears to be only slightly affected by temperature. For poly(siloxanes) containing bulky or polar functional groups, the regular helical conformation of the polymers is distorted resulting in a greater change in viscosity with temperature. These phases form films that are not as resistant to temperature variations as the poly(dimethylsiloxanes) but are generally significantly more stable than other types of common liquid phases.

A second important breakthrough in open tubular column technology was the realization that crosslinking of gum phases to form a rubber provided a means of further stabilizing the poly(siloxane) films without destroying their favorable diffusion characteristics. The unmatched flexibility of the silicon oxygen bond imparts great mobility to the polymer chains, providing openings that permit diffusion, even for crosslinked phases. Modern stationary phases used to prepare open tubular columns are characterized by high viscosity, good diffusivity, low glass transition temperatures and are often suitable for crosslinking. For the present suitable properties are only apparent for the poly(siloxane) and poly(ethylene glycol) phases, Table 2.5, explaining how

Table 2.5
Rough guide to the temperature operating range for bonded poly(siloxane) stationary phases in open tubular columns

| Type | Temperature Range (°C) Minimum | Maximum | High Temperature Version |
|---|---|---|---|
| Dimethylsiloxane | -60 | 325 | 420 |
| Dimethyldiphenylsiloxane (5 % diphenyl) | -60 | 325 | 420 |
| Dimethyldiphenylsiloxane (35 % diphenyl) | 40 | 300 | 340 |
| Dimethyldiphenylsiloxane (50 % diphenyl) | 40 | 325 | 390 |
| Methylphenylsiloxane | 0 | 280 | |
| Dimethyldiphenylsiloxane (65 % diphenyl) | 50 | 260 | 370 |
| 3,3,3-Trifluoropropylmethylsiloxane (50 % trifluoropropyl) | 45 | 240 | 300 |
| 3-Cyanopropylphenyldimethylsiloxane (6 % cyanopropylphenyl and 84 % dimethyl) | 20 | 280 | |
| 3-Cyanopropylphenyldimethylsiloxane (25 % cyanopropylphenyl and 75% dimethyl) | 40 | 240 | |
| 3-Cyanopropylphenyldimethylsiloxane (50 % cyanopropylphenyl and 50 % dimethyl) | 40 | 230 | |
| 3-cyanopropyl-silphenylene co-polymer (equivalent to 70 % dicyanopropyl) | | | 290 |
| Poly(ethylene glycol) | 20 | 250 | 280 |
| FFAP | 40 | 250 | |

they have come to dominate the practice of open tubular column gas chromatography, even more so than the poly(siloxane) phases dominate the field of packed column gas chromatography.

Two different approaches have been used to immobilize poly(siloxane) phases for the preparation of open tubular columns. Thermal immobilization of silanol-terminated non-polar and moderately polar poly(siloxanes) [73,74,84-86] and radical initiated crosslinking of endcapped poly(siloxanes) of a wider polarity range [62,78-80]. Carboxyalkyl-functionalized poly(siloxane) copolymers are also readily immobilized by thermal treatment but are not considered here [87]. For thermal condensation the acid leached and perhaps deactivated glass surface is statically coated with a film of the silanol-terminated polymer (occasionally methoxy-terminated polymers are also used). The column is then heated to 300-370°C for various times (5 to 15 h) in the presence of a slow flow of carrier gas. Thermal induced reactions result in simultaneous bonding of the polymer to the column wall and the formation of crosslinks between polymer chains. Addition of a few percent of a crosslinking agent, such as an alkyltrimethoxysilane, is sometimes used to improve the thermal stability of the film by promoting additional crosslinking and surface bonding.

The crosslinking reaction indicated above is important for another reason. It is also the root cause of thermal degradation of poly(siloxanes) resulting in column bleed at high temperatures. The main degradation products from the heating of silanol-terminated poly(siloxanes) have been identified as hydrocarbons and small ring cyclic siloxanes [72,84,85]. The formation of hydrocarbons is accounted for by a reaction leading to chain branching identical to that responsible for crosslinking in the column preparation step. The silanol group of one polymer chain acting as a nucleophile to displace the alkyl group from a neighboring polymer chain with the joining of the two chains by a new siloxane bond (crosslink). Similarly, reaction of the terminal silanol groups with surface-bonded organosiloxane groups (formed during deactivation by high temperature silanization) results in covalent bonding of the phases to the column wall. On undeactivated glass surfaces it is probably the reaction of silanol groups on the glass surface with siloxane groups of the polymer backbone that is responsible for surface bonding rather than condensation of polymer and surface silanol groups. Above some critical temperature the generation of crosslinks will be too high and the resulting film will have inadequate solvation and diffusion properties to be useful for gas chromatography. Cyclics are formed by an intramolecular displacement in which the silanol end group attacks a silicon atom in the same chain setting free small cyclic siloxanes with formation of a new silanol group. The cyclic siloxanes released are mainly tri- and tetrameric fragments easily identified by mass spectrometry in the column bleed profile of poly(siloxane) phases. Poly(siloxanes) containing electron-withdrawing cyano groups on $\alpha$-alkylcarbon or fluorine on a $\alpha$- or $\beta$-alkylcarbon atoms have low thermal stability and are less useful for gas chromatography than substituents with an additional methylene group in the alkyl chain between silicon and the polar functional group. The replacement of a methyl group by phenyl increases the thermal stability of the poly(siloxanes). This results from a strengthening of the siloxane bond by increased participation of the lone pair electrons on oxygen but the silicon aryl bond is more polarized and should be easier to cleave in chemical reactions, such as those leading to crosslinking. Poly(siloxanes) containing 3-cyanopropyl groups are more thermally stable when a phenyl group is attached to the same silicon atom than when a methyl group is present.

The use of silanol-terminated poly(siloxanes) have been instrumental in the development of columns for high temperature gas chromatography, Table 2.5 [88-93]. This requires phases of exceptional thermal stability and starts with meticulous care in the preparation of the glass surfaces to minimize catalytic decomposition of the poly(siloxanes). High molecular weight polymers optimize the number of silanol groups per unit volume of polymer to produce the desired number of crosslinks and minimize the formation of cyclic siloxanes. The thermal stability of the poly(siloxanes) is improved if phenyl groups (silarylene copolymers) or carborane groups are incorporated into the poly(siloxane) backbone (Figure 2.2), the substitution on the poly(siloxane) backbone is symmetrical, and the sequence of silyl groups along the siloxane backbone is alternating. For example, a poly(methylphenylsiloxane) containing dimethylsiloxane and diphenylsiloxane units is likely to be more stable than a polymer containing methylphenylsiloxane units in its structure. Incorporating phenyl or carborane groups in the polymer backbone inhibits the formation of cyclic siloxanes by reducing the flexibility of the polymer chains. An example of the use of high temperature gas chromatography to separate hydrocarbons with a carbon number greater than 100 using tempera-

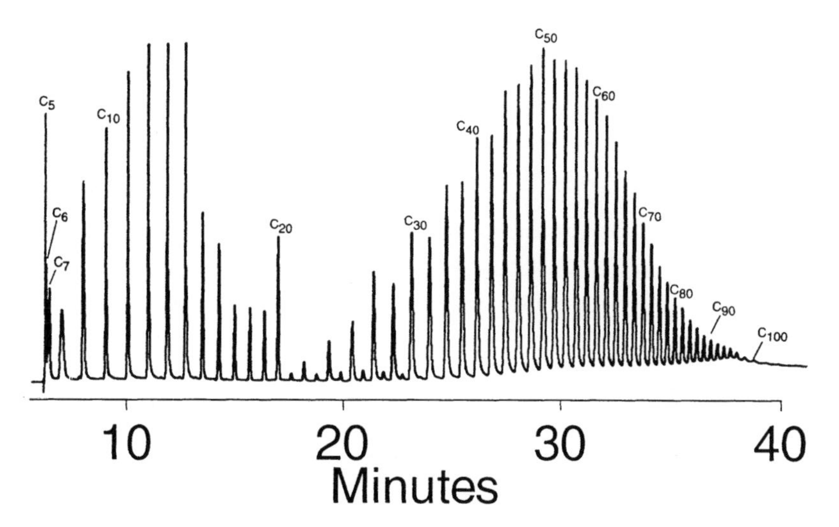


Figure 2.3. Separation of polywax 655 by high temperature gas chromatography on a 6 m x 0.53 mm I. D. open tubular column coated with a 0.1 μm film of a poly(carborane-siloxane) copolymer (equivalent to 5 % phenyl). Initial column temperature -20°C for 1 min, programmed at 10°C/min to 430°C, and final hold 5 min at 430°C. The helium carrier gas flow rate was 20 ml/min. (©SGE, Inc.)

ture programming up to a final temperature of 420°C is presented in Figure 2.3. High temperature gas chromatography is an important technique for the structural analysis of hydrocarbons and triglycerides of high carbon number [88,93-95].

The second general approach to immobilization of poly(siloxanes) is by free radical crosslinking of the polymer chains using peroxides [96-98], azo-compounds [80,97,99] or γ-radiation [100,101] as free radical generators. In this case, crosslinking occurs through the formation of carbon-carbon bonds involving the organic substituents on neighboring poly(siloxane) chains. The silicon atoms are not involved in the crosslinking reaction and no new siloxane bonds are formed. Very little crosslinking (0.1-1.0%) is required to immobilize long chain poly(siloxanes). Dicumyl peroxide and azo-*tert*-butane have emerged as the most widely used free radical generators for stationary phase immobilization.

Free radical crosslinking provides a relatively simple column preparation procedure. The deactivated glass surface is statically coated with a freshly prepared solution of the stationary phase containing 0.2 to 5% w/w of dicumyl peroxide or the coated column is saturated with azo-tert.-butane vapors entrained in nitrogen gas at room temperature for about 1 h. It is very important that the stationary phase is deposited as an even film since this film will be fixed in position upon crosslinking and no improvement in film homogeneity can be obtained after fixation. The coated column is sealed and slowly raised to the curing temperature for static curing. The curing temperature is a function of the thermal stability of the free radical generator and is selected to give a reasonable half-life for the crosslinking reaction. Once the column reaches the curing temperature, maintaining that temperature for a short time is all that is required to complete the reaction. Dynamic curing with a slow flow of carrier gas is sometimes used with dicumyl peroxide. After curing, any phase that is not immobilized is rinsed from the column by flushing with appropriate organic solvents. Poly(dimethylsiloxane) phases are relatively easy to immobilize by free radical crosslinking, but with increasing substitution of methyl by bulky or polar functional groups the difficulty of obtaining complete immobilization increases. For this reason moderately polar poly(siloxane) phases are prepared with small amounts of vinyl, tolyl or octyl groups attached to the siloxane backbone or as endcapping groups to increase the success of the crosslinking reaction. Even so, the complete immobilization of poly(cyanopropylsiloxane) phases with a high percentage of cyanopropyl groups has proven difficult. Commercially available immobilized poly(cyanopropylsiloxane) phases are either completely immobilized and contain less than about 50% substitution with cyanopropyl groups or they are "stabilized", indicating incomplete immobilization for phases with a higher incorporation of cyanopropyl groups. Stabilized columns usually have greater thermal stability than physically coated columns but are not resistant to solvent rinsing or as durable as immobilized columns.

Another general advantage of immobilization is that it allows the preparation of thick-film open tubular columns. It is very difficult to prepare conventionally coated columns with stable films thicker than about 0.5 μm. Columns with immobilized phases and film thicknesses of 1.0-8.0 μm are easily prepared. Thick-film columns allow the analysis of volatile substances with reasonable retention factors without the need for subambient temperatures. Immobilization also improves the resistance of the stationary phase to phase stripping by large volume splitless or on-column injection and allows solvent rinsing to be used to free the column from non-volatile sample byproducts or from active breakdown products of the liquid phase. Immobilization procedures also provide a useful increase in the upper column operating temperature limit. Thus, immobilization techniques were important not only in advancing column technology but also contributed to advances in the practice of gas chromatography by facilitating injection, detection and other instrumental developments that would have been difficult without them.

### 2.3.5 Solvation Properties of Stationary Phases

Solvent strength and selectivity are the properties commonly used to classify liquid stationary phases as selection tools for method development in gas chromatography [29,102-104]. Solvent strength and polarity are often used interchangeably and can cause confusion. Polarity is sometimes considered to be the capacity of a stationary phase for dipole-type interactions alone, while more generally solvent strength is defined as the capacity of a stationary phase for all possible intermolecular interactions. The latter definition is quite sensible but unworkable because there is no substance that is uniquely polar that might be used to probe the polarity of other substances. Indirect measurements of polarity, such as those scales related one way or another to the reluctance of a polar stationary phase to dissolve a methylene group (or n-alkane) can only determine the capacity of the stationary phase for non polar interactions (dispersion and induction). It can never adequately characterize the capacity of a solvent for specific interactions such as orientation and hydrogen bonding, although these interactions contribute to the solvation process of a methylene group in the form of the greater free energy required for cavity formation. A strong hydrogen-bond acid stationary phase with weak hydrogen-bond basicity, for example, will undergo weaker interactions with itself (solvent-solvent interactions) than with a solute that is a strong hydrogen-bond base. In this case the solvation process for a methylene group will not adequately characterize the capacity of the stationary phase for hydrogen bond formation. Because of these considerations and others it is necessary to abandon the use of stationary phase classification based on a single parameter, indicated as polarity, in spite of the obvious emotive attributes of such a simple scale.

The selectivity of a stationary phase is defined as its relative capacity for specific intermolecular interactions, such as dispersion, induction, orientation and complexation (including hydrogen bond formation). Unlike solvent strength (polarity) it should be feasible to devise experimental scales of stationary phase selectivity. Early attempts to define selectivity scales were based on the system of phase constants introduced by Rohrschneider and subsequently modified by McReynolds (section 2.4.4.2), Snyder's solvent selectivity triangle [105], Hawkes polarity indices [106], solubility parameters [107], and the Gibbs free energy of solution for selected solutes or functional groups [39,108]. These approaches are reviewed in detail elsewhere [28-30,102,103,109,110], but are no longer relevant for stationary phase classification.

Modern approaches to stationary phase classification by selectivity ranking are based on the cavity model of solvation [102,104,111,112]. This model assumes that the transfer of a solute from the gas phase to solution in the stationary phase involves three steps. Initially a cavity is formed in the stationary phase identical in volume to the solute. The solute is then transferred to the cavity with reorganization of the solvent molecules around the cavity and the set up of solute-solvent interactions. The individual free energy terms involved in the transfer are assumed to be additive. In addition, it can be assumed that the free energy associated with cavity reorganization is small compared to the other processes and that the gas-solute vapors behave ideally. Consequently, gas-solute and solute-solute interactions in the gas phase are negligible. For separation conditions employed in analytical gas chromatography infinite dilution conditions can be assumed so that all solution interactions are of the solvent-solvent and solute-solvent type. These interactions can be characterized as dispersion, induction, orientation and complexation. Dispersion interactions are non-selective and are the binding forces that hold all molecular assemblies together. Molecules with a permanent dipole moment can interact with each other by the co-operative alignment of their dipoles (orientation interactions) and by their capacity to induce a temporary complementary dipole in a polarizable molecule (induction interactions). Complexation interactions are selective interactions involving the sharing of electron density or a hydrogen atom between molecules (e.g. hydrogen bonding and charge transfer). In gas-liquid chromatography retention will depend on the cohesive energy of the stationary phase represented by the free energy required for cavity formation in the stationary phase, the formation of additional dispersion interactions of a solute-solvent type, and on selective solute-solvent polar interactions dependent on the complementary character of the polar properties of the solute and solvent.

The master retention equation of the solvation parameter model relating the above processes to experimentally quantifiable contributions from all possible intermolecular interactions was presented in section 1.4.3. The system constants in the model (see Eq. 1.7 or 1.7a) convey all information of the ability of the stationary phase to participate in solute-solvent intermolecular interactions. The $r$ constant refers to the ability of the stationary phase to interact with solute n- or $\pi$-electron pairs. The $s$ constant establishes the ability of the stationary phase to take part in dipole-type interactions. The $a$ constant is a measure of stationary phase hydrogen-bond basicity and the $b$ constant stationary phase hydrogen-bond acidity. The $l$ constant incorporates contributions from stationary phase cavity formation and solute-solvent dispersion interactions. The system constants for some common packed column stationary phases are summarized in Table 2.6 [68,81,103,104,113]. Further values for non-ionic stationary phases [114,115], liquid organic salts [68,116], cyclodextrins [117], and lanthanide chelates dissolved in a poly(dimethylsiloxane) [118] are summarized elsewhere.

The system constants in Table 2.6 are only loosely scaled to each other so that changes in magnitude in any column can be read directly but changes in magnitude along rows must be interpreted cautiously. Most stationary phases possess some capacity for electron lone pair interactions ($r$ constant), but selectivity for this interaction is rather limited among common stationary phases. Fluorine-containing stationary phases have negative values of the $r$ constant representing the tighter binding of electron pairs in fluorocarbon compared to hydrocarbon groups. Electron lone pair interactions do not usually make a significant contribution to retention in gas-liquid chromatography and are not considered as a primary means of selectivity optimization. The most striking feature of Table 2.6 is the paucity of stationary phases with significant hydrogen-bond acidity ($b$ constant). In the case of EGAD, DEGS and TCEP the small $b$ constant is more likely a reflection of impurities in the stationary phase produced during synthesis or while in use [113]. Many stationary phases contain hydrogen-bond acid groups such as hydroxyl, amide or phenol groups that are expected to behave as hydrogen-bond acids. These groups are also significant hydrogen-bond bases and prefer to self-associate forming inter- and intramolecular hydrogen-bond complexes to the exclusion of hydrogen-bond acid interactions with basic solutes. It is only recently that suitable stationary phases with hydrogen-bond acid properties have been described, Figure 2.4 [81,119,120]. Outstanding among these is PSF6 which has a wide temperature operating range and is a strong hydrogen-bond acid with zero hydrogen-bond basicity. These properties reflect the chemistry of the fluorinated alcohol portion of the poly(siloxane) polymer. The strong electron withdrawing trifluoromethyl groups lower the electron density on the neighboring oxygen atom of the hydroxyl group increasing its hydrogen-bond acidity while simultaneously reducing its hydrogen-bond

Table 2.6
System constants derived from the solvation parameter model for packed column stationary phases at 121°C

| Stationary phase | System constant | | | | | |
|---|---|---|---|---|---|---|
| | *r* | *s* | *a* | *b* | *l* | *c* |
| *(i) Hydrocarbon phases* | | | | | | |
| Squalane | 0.129 | 0.011 | 0 | 0 | 0.583 | -0.222 |
| Apolane-87 | 0.170 | 0 | 0 | 0 | 0.549 | -0.221 |
| *(ii) Ether and ester phases* | | | | | | |
| Poly(phenyl ether) 5 rings PPE-5 | 0.230 | 0.829 | 0.337 | 0 | 0.527 | -0.395 |
| Carbowax 20M CW20M | 0.317 | 1.256 | 1.883 | 0 | 0.447 | -0.560 |
| Poly(ethylene glycol) Ucon 50 HB 660 | 0.372 | 0.632 | 1.277 | 0 | 0.499 | -0.184 |
| 1,2,3-Tris(2-cyanoethoxypropane) TCEP | 0.116 | 2.088 | 2.095 | 0.261 | 0.370 | -0.744 |
| Didecylphthalate DDP | 0 | 0.748 | 0.765 | 0 | 0.560 | -0.328 |
| Poly(ethylene glycol adipate) EGAD | 0.132 | 1.394 | 1.820 | 0.206 | 0.429 | -0.688 |
| Poly(diethylene glycol succinate) DEGS | 0.230 | 1.572 | 2.105 | 0.171 | 0.407 | -0.650 |
| *(iii) Liquid organic salts* | | | | | | |
| Tetrabutylammonium 4-toluenesulfonate QBApTS | 0.156 | 1.582 | 3.295 | 0 | 0.459 | -0.686 |
| Tetrabutylammonium tris(hydroxymethyl)methyl-amino-2-hydroxy-1-propanesulfonate QBATAPSO | 0.266 | 1.959 | 3.058 | 0 | 0.317 | -0.860 |
| Tetrabutylammonium 4-morpholinepropane-sulfonate QBAMPS | 0 | 1.748 | 3.538 | 0 | 0.550 | -0.937 |
| Tetrabutylammonium methanesulfonate QBAMES | 0.334 | 1.454 | 3.762 | 0 | 0.435 | -0.612 |
| *(iv) Poly(siloxane) phases* | | | | | | |
| Poly(dimethylsiloxane) SE-30 | 0.024 | 0.190 | 0.125 | 0 | 0.498 | -0.194 |
| Poly(dimethylmethylphenylsiloxane) OV-3 (10 mol % phenyl) | 0.033 | 0.328 | 0.152 | 0 | 0.503 | -0.181 |
| Poly(dimethylmethylphenylsiloxane) OV-7 (20 mol % phenyl) | 0.056 | 0.433 | 0.165 | 0 | 0.510 | -0.231 |
| Poly(dimethylmethylphenylsiloxane) OV-11 (35 mol % phenyl) | 0.097 | 0.544 | 0.174 | 0 | 0.516 | -0.303 |
| Poly(methylphenylsiloxane) OV-17 | 0.071 | 0.653 | 0.263 | 0 | 0.518 | -0.372 |
| Poly(methylphenyldiphenylsiloxane) OV-22 (65 mol % phenyl) | 0.201 | 0.664 | 0.190 | 0 | 0.482 | -0.328 |
| Poly(methylphenyldiphenylsiloxane) OV-25 (75 mol % phenyl) | 0.277 | 0.644 | 0.182 | 0 | 0.472 | -0.273 |
| Poly(cyanopropylmethyldimethylsiloxane) (10 mol % cyanopropylmethylsiloxane) OV-105 | 0 | 0.364 | 0.407 | 0 | 0.496 | -0.203 |
| Poly(cyanopropylmethylphenylmethylsiloxane) (50 mol % cyanopropylmethylsiloxane) OV-225 | 0 | 1.226 | 1.065 | 0 | 0.466 | -0.541 |
| Poly(dicyanoalkylsiloxane) OV-275 (70 mol % dicyanopropyl and 30 mol % dicyanoethyl) | 0.206 | 2.080 | 1.986 | 0 | 0.294 | -0.909 |
| Poly(trifluoropropylmethylsiloxane) QF-1 | -0.449 | 1.157 | 0.187 | 0 | 0.419 | -0.269 |
| Poly(dimethylsiloxane)-Poly(ethylene glycol) Copolymer OV-330 | 0.104 | 1.056 | 1.419 | 0 | 0.481 | -0.430 |
| PSF6 (see Fig. 2.4) | -0.360 | 0.820 | 0 | 1.110 | 0.540 | -0.510 |
| *(v) Miscellaneous* | | | | | | |
| Bis(3-allyl-4-hydroxyphenyl)sulfone H10 | -0.051 | 1.323 | 1.266 | 1.457 | 0.418 | -0.568 |

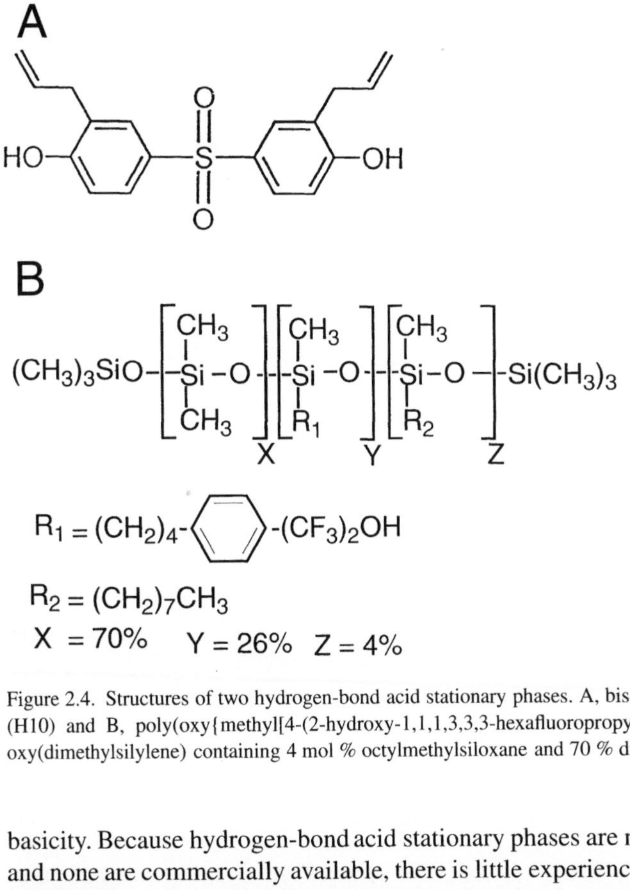

A

B

$R_1 = (CH_2)_4$-[phenyl]-$(CF_3)_2OH$

$R_2 = (CH_2)_7CH_3$

X = 70% Y = 26% Z = 4%

Figure 2.4. Structures of two hydrogen-bond acid stationary phases. A, bis(3-allyl-4-hydroxyphenyl)sulfone (H10) and B, poly(oxy{methyl[4-(2-hydroxy-1,1,1,3,3,3-hexafluoropropyl-2-yl)phenyl]butyl}silylene)-*co*-oxy(dimethylsilylene) containing 4 mol % octylmethylsiloxane and 70 % dimethylsiloxane (PSF6).

basicity. Because hydrogen-bond acid stationary phases are new to gas chromatography and none are commercially available, there is little experience in their use for selectivity optimization. That leaves the most important stationary phase properties for selectivity optimization as their cohesive energy and capacity for dipole-type and hydrogen-bond base interactions.

The solvation parameter model cannot separate the cohesive solvent-solvent and solute-solvent dispersion interactions since both are strongly correlated to the solute volume. In the general form of the solvation parameter model they are contained in the contribution of the model constant ($c$ term) and in the $l$ term. The model constant term contains other factors besides a contribution from cavity formation, including the phase ratio when the dependent variable is other than the gas-liquid distribution constant (e.g. the retention factor), scaling errors in the solute descriptors, and contributions of a statistical nature from a lack of fit to the model. Proceeding cautiously we can use the product term $\Sigma(c + l \log L^{16})$ to assess the importance of the cavity/dispersion term to retention. Using decane as an example, some data for $\Sigma(c + l \log L^{16})$ are summarized in Table 2.7. In all cases the product term $\Sigma(c + l \log L^{16})$ is positive indicating that the solute-solvent dispersion interactions setup by placing decane into

Table 2.7
Contribution of cavity formation and dispersion interactions to solution of decane in different stationary phases at 121°C
(QBATS = tetrabutylammonium 4-toluenesulfate).

| Stationary phase | $c$ | $l \log L^{16}$ | $\Sigma(c + l \log L^{16})$ |
|---|---|---|---|
| Squalane | -0.222 | 2.732 | 2.510 |
| SE-30 | -0.194 | 2.334 | 2.140 |
| PPE-5 | -0.395 | 2.470 | 2.075 |
| OV-17 | -0.372 | 2.427 | 2.055 |
| QF-1 | -0.269 | 1.963 | 1.694 |
| OV-225 | -0.541 | 2.184 | 1.643 |
| Carbowax 20M | -0.560 | 2.095 | 1.535 |
| QBATS | -0.686 | 2.151 | 1.465 |
| DEGS | -0.650 | 1.907 | 1.257 |
| TCEP | -0.744 | 1.734 | 0.990 |
| OV-275 | -0.909 | 1.378 | 0.469 |

the cavity exceed the free energy required to disrupt the solvent structure in forming the cavity. The cavity/dispersion term for the hydrocarbons, poly(dimethylsiloxanes) and poly(methylphenylsiloxanes) are in fact all very similar indicating roughly equal difficulty in forming a cavity in the stationary phase [121]. Stationary phases with dipolar and hydrogen bonding functional groups are considerably more cohesive and the additional free energy required for cavity formation is reflected in the significantly smaller value for $\Sigma(c + l \log L^{16})$. This is particularly so for the poly(cyanoalkylsiloxane) stationary phase OV-275, for which the product term $\Sigma(c + l \log L^{16})$ is considerably smaller than for all other phases in Table 2.7, and the high cohesive energy of this phase is a significant factor in explaining its selectivity. From an interpretive point of view the $l$ constant indicates the spacing between members of a homologous series and contains useful information for phase selection. There is generally a good correlation between the $l$ system constant and the partial molar Gibbs free energy of solution for a methylene group. The liquid organic salts with non-associating anions have surprisingly large $l$ constants compared to non-ionic polar stationary phases, 0.44 to 0.55, and are unique among polar stationary phases in their ability to separate compounds belonging to a homologous series [68]. For anions believed to be associated as hydrogen-bond complexes the $l$ constants are significantly smaller, 0.26 to 0.37, and equivalent to the values observed for the most polar non-ionic stationary phases. An example of the unique selectivity and modest cohesive energy of a non-associated liquid organic salt, tetrabutylammonium 4-toluenesulfonate to separate a test mixture of polar compounds compared to a polar non-ionic phase, OV-275, is shown in Figure 2.5.

The stationary phases in Table 2.6 differ significantly in their capacity for dipole-type interactions ($s$ constant) and in their hydrogen-bond basicity ($a$ constant). For the poly(methylphenylsiloxanes) increasing the phenyl content up to 50 mol % phenyl groups produces an orderly change in the capacity of the phases for dipole-type interactions and a shallow change in their capacity for interactions as a hydrogen-

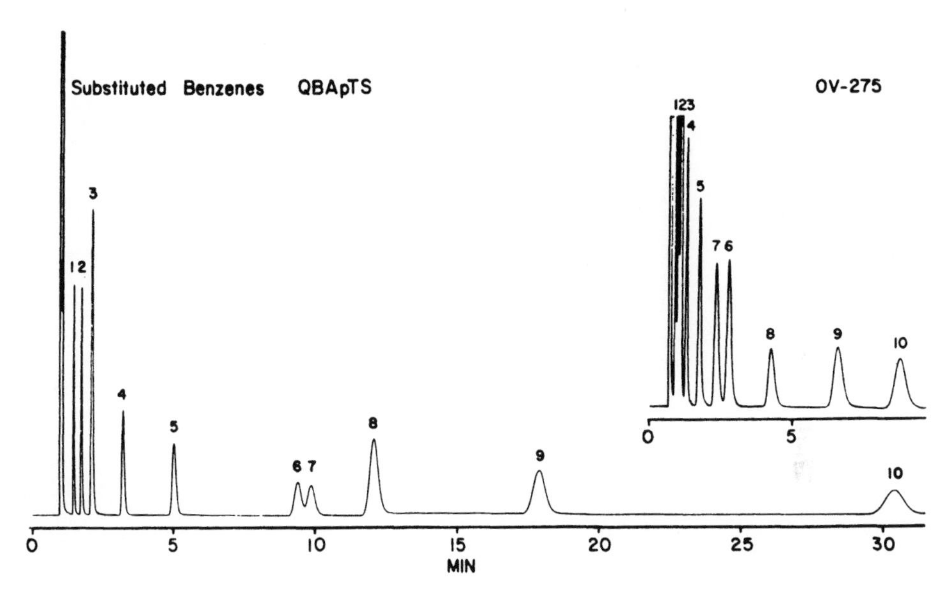


Figure 2.5. Separation of a mixture of polar compounds on matched packed columns coated with tetrabutylammonium 4-toluenesulfonate (QBApTS) and OV-275. Each column was 3.5 m x 2 mm I.D. containing 10% (w/w) of stationary phase on Chromosorb W-AW (100-120 mesh) with a carrier gas flow rate of 15 ml/min and column temperature 140°C. Peak assignments: 1 = benzene; 2 = toluene; 3 = ethylbenzene; 4 = chlorobenzene; 5 = bromobenzene; 6 = iodobenzene; 7 = 1,2-dichlorobenzene; 8 = benzaldehyde; 9 = acetophenone; and 10 = nitrobenzene.

bond base, Figure 2.6. Above 50 mol % phenyl groups there is a slight decline in the system constants as additional phenyl groups are added as diphenylsiloxane monomer units. The $r$ system constant, which is numerically small up to 50 mol % phenyl groups, shows an abrupt increase in value as diphenylsiloxane groups are introduced into the poly(siloxane). Given that the cavity/dispersion term of the poly(methylphenylsiloxanes) changes little with composition, the principal selectivity difference among these phases is due to changes in their capacity for dipole-type interactions. Above 50 mol % phenyl groups selectivity differences are small and result primarily from changes in cohesive energy and lone-pair electron interactions.

Principal component analysis and hierarchical clustering methods provide useful approaches for stationary phase classification [68,81,103,1113,114,121]. The results from principal component analysis for 52 varied non hydrogen-bond acidic stationary phases with their system constants entered as variables are shown in Figure 2.7. The first two principal components account for about 95% of the total variance with the first principal component (PC 1) strongly associated with hydrogen-bond base interactions ($a$ constant) and the second principal component (PC 2) strongly associated with dipole-type interactions and phase cohesive energy ($s$ and $l$ system constants). The stationary phases are classified into three disperse groups. The group 1 stationary phases (e.g.

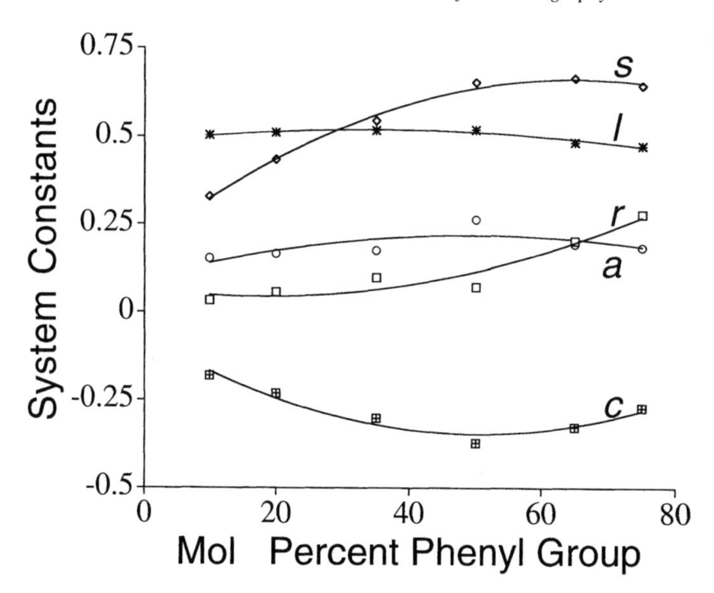


Figure 2.6. Plot of the system constants against the mol % phenyl composition for a series of poly(methylphenylsiloxane) and poly(methylphenyldiphenylsiloxane) phases.

squalane, SE-30, OV-3, OV-7, OV-105, OV-11, OV-17, OV-22, OV-25, PPE-5, QF-1) are weak hydrogen-bond bases with a weak and variable capacity for dipole-type interactions. Differences in cohesive energy are small. Group 2 contains the polar and cohesive non-ionic stationary phases and some liquid organic salts with highly delocalized anions (e.g. U50HB, OV-225, OV-275, OV-330, TCEP, CW 20M, EGAD, DEGS, tetrabutylammonium picrate). These phases have a narrow range of hydrogen-bond basicity and are differentiated primarily by differences in their capacity for dipole-type interactions and their cohesive energy. Group 3 contains only liquid organic salts, which are all dipolar ($s$ = 1.4 to 2.1) and strong hydrogen-bond bases ($a$ = 1.4 to 5.4) with variable cohesive energy. The non-ionic stationary phases have $s$ and $a$ system constants that each cover the same range of 0 to 2.1. The hydrogen-bond basicity of the liquid organic salts depends primarily on the extent of charge delocalization and anion size. Charge delocalizing anions (e.g. picrate, perfluorobenzenesulfonate) are weaker hydrogen-bond bases and less dipolar than the other salts. Small non-delocalizing anions (e.g. chloride, bromide) are the most hydrogen-bond basic. Anions containing hydrogen bonding substituents have only a small influence on the dipolarity and hydrogen-bond basicity of the salts ($s$ and $a$ constants only slightly changed) but result in a significant increase in their cohesive energy ($l$ constant reduced).

Cluster analysis is an alternative to principal component analysis for the classification of samples by multivariate analysis. The outputs for clustering algorithms are dendrograms. The complete link dendrogram for the stationary phases listed in Table 2.6 is shown in Figure 2.8. The stationary phases most similar to each other are next to each

```
Top-4 { --clearinterrupt-- --disableinterrupt-- (   )
        --exch-- 0 --exch-- --put-- --clear-- }
```

**ctionary Stack**

```
Top    -dict-
Top-1 -dict-
Top-2 -dict-
Top-3 -dict-
Top-4 -dict-
```

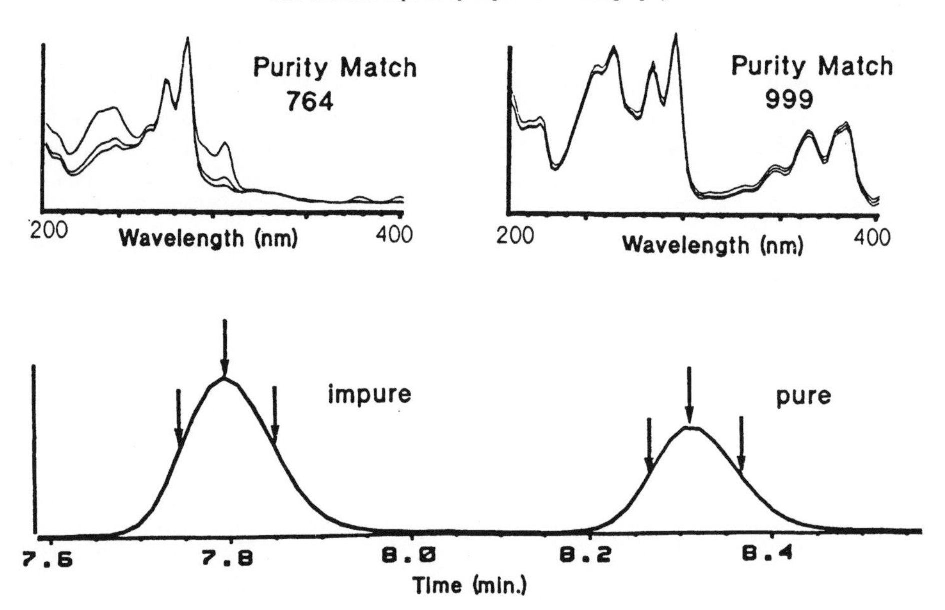


Figure 5.12. Example of peak purity determination by matching spectra recorded at different peak positions. (From ref. [107]; ©Marcel Dekker).

Diode array detectors are also used for peak purity and homogeneity analysis and peak identification. One common method of peak purity analysis is based on the use of absorbance ratios [119]. The computation of the absorbance ratio at two wavelengths for all points across the peak profile is a square-wave function for a single-component peak. Coeluting impurities, which absorb at least one of the selected wavelengths, will distort this function. The selection of detection wavelengths and the absorption threshold value can dramatically affect the utility of this technique. A superior approach is the comparison of the various spectra recorded during the elution of the peak to either a reference spectra or to each other. If an impurity has roughly the same peak shape as the main component but elutes at a slightly different retention time, then the level of the impurity will not be constant across the peak. Differences are expected to be greatest at the beginning and end of the main peak but the extreme regions of the main peak are most affected by noise and therefore less useful. A more satisfactory solution is the comparison of spectra recorded on the up slope, apex and down slope of the main peak, Figure 5.12 [107]. Irrespective of the actual way in which spectra are compared, an impurity is identified as a deviation in the match factor from the ideal value for identical spectra by an amount greater than can be attributed to noise. Even if the spectra are identical a peak can still be impure due to any of the following reasons: the impurity is present at a low concentration and therefore not detectable; the impurity has the same or very similar spectrum to the main component; or the impurity has the same peak profile and elutes at the same time as the main peak.

For overlapping peaks the data matrix contains linear combinations of the pure spectra of the overlapping components in its rows, and combinations of the pure elution profiles in its columns. Multivariate analysis of the data matrix may allow extraction of useful information from either the rows or columns of the matrix, or an edited form of the data matrix [107,116-118]. Factor analysis approaches or partial least-squares analysis can provide information on whether a given spectrum (known compound) or several known compounds are present in a peak. Principal component analysis and factor analysis can be used to estimate the maximum number of probable (unknown) components in a peak cluster. Deconvolution or iterative target factor analysis can then be used to estimate the relative concentration of each component with known spectra in a peak cluster.

Peak identification is based on the comparison of normalized spectra representative for the peak with spectra of one or several standard compounds run in the same separation system and stored in a spectral library [107,116]. This approach is less powerful than for mass or infrared spectral searches due to the rather broad and featureless bands that typify absorption spectra. Absorption spectra of similar compounds and compounds with a chromophore well separated from the variation in molecular structure are often virtually identical. Also, spectral changes dependent on the experimental conditions (pH, mobile phase composition, temperature, etc.) occur frequently. For this reason user prepared local libraries tend to predominate over general libraries, in contrast to common practices in infrared and mass spectral searches. A favorable spectral match for an absorption spectrum by itself is not acceptable for absolute identification.

Spectral matching is the process used to establish the similarity of compared spectra as a single number. Most matching procedures employ a point by point comparison of the two spectra in digital format to establish the presence or absence of significant differences. Individual matching procedures differ primarily in the methods used for spectral normalization and subsequently in the way spectral differences between spectra are converted to a single number. One common approach, for example, views each spectrum as a vector in multidimensional space and uses the angle between vectors (spectra) as a measure of their similarity. Library searches are usually performed as a forward or reverse search. In forward searches an attempt is made to identify each compound in an unknown sample from a large library of standard compounds. In a reverse search a limited library of standard compounds, all of which are likely to be present in the sample type, is searched against the unknown spectra in the current separation. Library searches can be combined with retention information to improve the certainty of identification. This is done by either weighting the match factor and retention similarity or by using the retention time to establish a retention window to preselect candidate spectra for comparison to the unknown spectra. Spectral quality is often an issue when matrix interference or noise precludes accurate spectral recording.

Careful consideration must be given to the design of the detector flow cell as it forms an integral part of both the chromatographic and optical systems. A compromise between the need to miniaturize the cell volume to reduce extracolumn band broadening

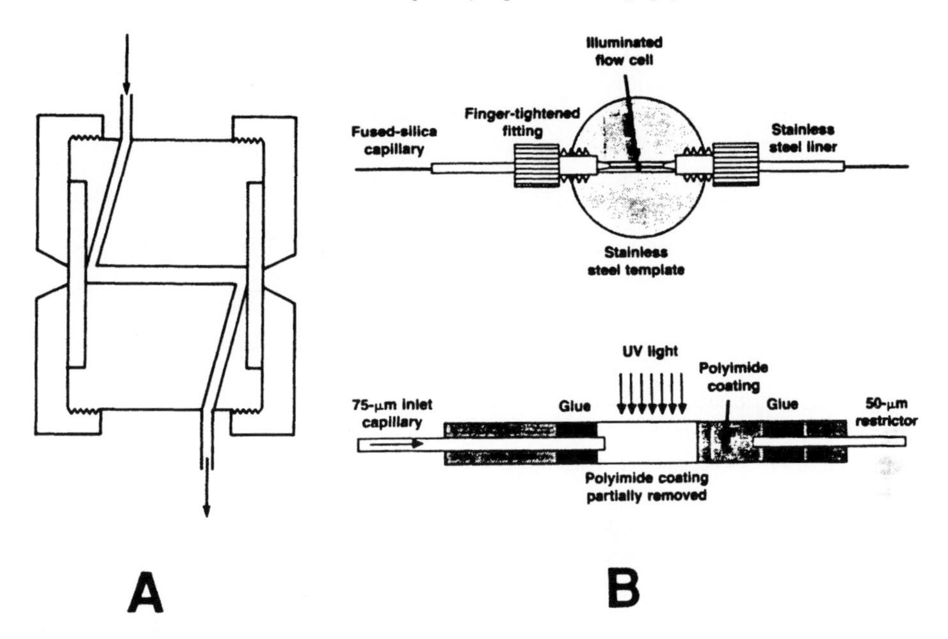


Figure 5.13. Representative flow cells used for absorption detection. A, Z-cell for conventional packed column applications and B, a low-volume fiber optic cell (volume = 4 nl and path length 0.15 mm) for use with packed capillary columns.

and the desire to employ long path lengths to increase sample detectability must be made. For packed columns of 4-8 mm internal diameter flow cells of the H-cell, Z-cell (Figure 5.13), or tapered cell configuration with a cell volume of about 5-10 μl and a path length of 1 cm, are commonly used [101,120]. The Z-cell design is used to minimize stagnant flow regions in the cell and to reduce peak tailing. Typically, the flow is confined by two quartz windows held in place by caps screwed onto the cell body at either end of the tubular cell cavity. The tapered cell has a conical cavity with a narrower aperture at the entrance than at the exit. The purpose of the tapered cavity is to reduce the amount of refracted light lost to the cell walls. Miniaturization of the same cell designs for use with small bore columns is possible by reducing the path length to 3 to 5 mm, providing cell volumes of about 0.2 to 2 μl. In order to maintain sensitivity the cell diameter should be reduced while maintaining the longest practical path length dictated by the cell volume requirements. Ultimately, however, reducing the cell diameter decreases the light throughput and increases background noise and susceptibility to refractive index changes. Commercially available flow cells suitable for absorption detection with small-bore columns have internal diameters of about 0.5 mm and may be internally coated with a reflective metallic layer to further reduce signal fluctuations caused by refractive index changes in the column eluent.

For absorption detection using packed capillary columns, detector volumes must be reduced to the nl range [109,114,121-124]. The simplest flow cells are prepared by removing a short length of the polyimide protective coating from a fused-silica capillary, just beyond the retaining frit for the column packing. The cell volume and path length depends on the internal diameter of the fused-silica capillary. The detector volume is also defined by the slit width in the orthogonal direction to the capillary diameter, which cannot be made too small, otherwise there will be very little light transmitted through the capillary. Cell volumes from about 3 to 200 nl are easily created in this way. A fixed fused-silica capillary configured into a U- or Z-shape mounted in a holder can be more convenient in practice, since different columns can be attached using a short piece of PEEK tubing without changing the optical alignment for the cell. Also, the cell path length can be made wider than the internal diameter of the separation capillary to increase sample detectability at the possible expense of additional band broadening due to the increase in capillary diameters. Axial illumination achieved by coupling the source to a fiber optic positioned at the exit of the fused-silica capillary has also been explored to increase light throughput and the cell path length. The small size and cylindrical walls of capillary cells present problems in focusing the source energy into capillary cells. If a significant portion of the light passes through the cell wall, but not the fluid sample stream (as can easily be the case, since the column wall is typically thicker than the internal diameter of the capillary), the linear response range for the detector will be significantly reduced. Extreme sensitivity to refractive index perturbations is also common for all small-volume cell designs. A novel small-volume flow cell constructed from a cross fitting and two optical fibers is shown in Figure 5.13 [107]. The cell has no windows and can be located away from the source and photosensor. The exit capillary is narrower than the column internal diameter to maintain sufficient backpressure to avoid outgassing of the mobile phase. This cell has a volume of about 4 nl and a path length of 0.15 mm.

The principal source of background noise in modern absorption detectors is attributed to the inhomogeneous changes in refractive index of the eluent flowing through the cell [120,124-128]. This arises from temperature gradients, incomplete mixing of the mobile phase, flow perturbations, and turbulence. These effects cause some of the incident light, which would normally pass directly through the cell, to strike the cell wall instead, and is lost. The photosensor cannot distinguish light lost by refraction to the cell walls from light absorbed by the sample. The continuous variation of the refracted light contribution to the detector signal constitutes the major portion of the noise signal observed when the detector is operated at high sensitivity. Employing an optical design with aperture and field stops external to the cell is generally used to minimize the contribution from variations in the refraction of the flow cell.

#### *5.7.1.2 Fluorescence Detectors*

Fluorescence detectors differ mainly in the choice of excitation source and the method used to isolate the excitation and emission wavelengths [101-106,108,129,130]. Since the signal intensity is directly proportional to the source intensity, high-energy line (mercury) or continuous (deuterium or xenon) arc sources are used. The mercury source produces line spectra (254, 313 and 365 nm) which can be isolated with

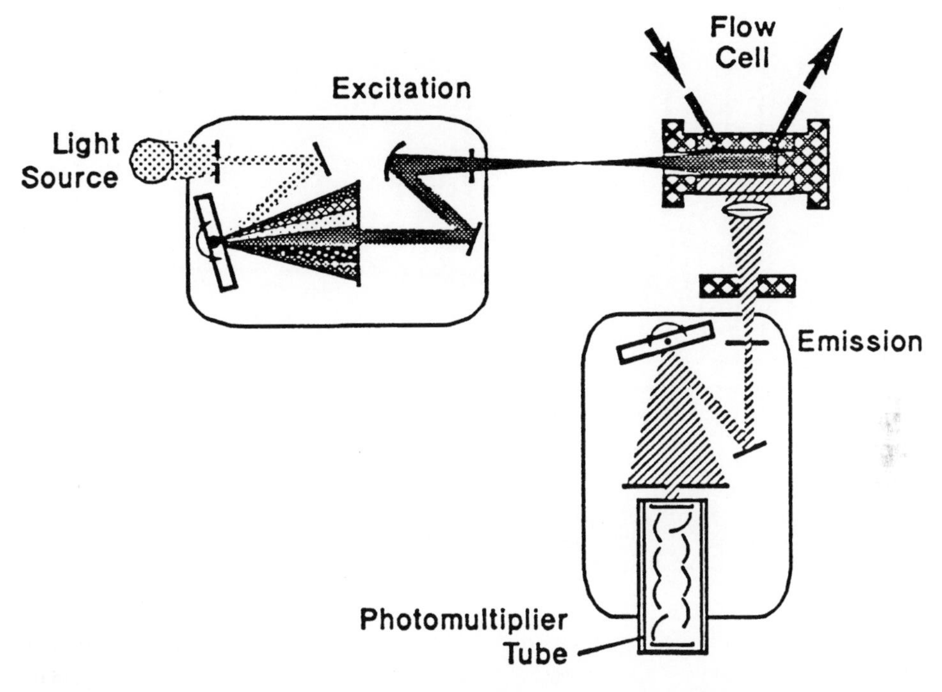


Figure 5.14. Schematic diagram of a fluorescence detector with rapid scanning monochromators for programmable selection of excitation and emission wavelengths.

a simple filter. However, the available emission lines may not overlap with the maximum excitation wavelength of the sample diminishing the detector response. Deuterium (190-400 nm) and xenon (200-850 nm) arc sources are used with a monochromator for continuously variable selection of the excitation wavelength, Figure 5.14. Fluorescence emission occurs in all directions and a hemispherical mirror can be used to maximize the fluorescence emission signal collected with an end-on photomultiplier detector. Emission wavelength isolation is performed using either filters or a monochromator. Filters are low cost and simple to use, however, there is generally a reduction in selectivity relative to wavelength selection using a monochromator because filters usually pass a wider range of wavelengths. Many modern detectors can be programmed to change either or both excitation and emission wavelengths during a separation to maintain optimum selectivity and sensitivity throughout the chromatogram, Figure 5.15. Some detectors using rapid scanning monochromators or a polychromator and an intensified diode array detector can record emission spectra at a fixed excitation wavelength as a function of retention time [108,130-133]. Less commonly rapid scanning monochromators or polychromators with diode array or charged coupled detectors have been used for the simultaneous recording of excitation and emission spectra as a function of retention time [119,129].

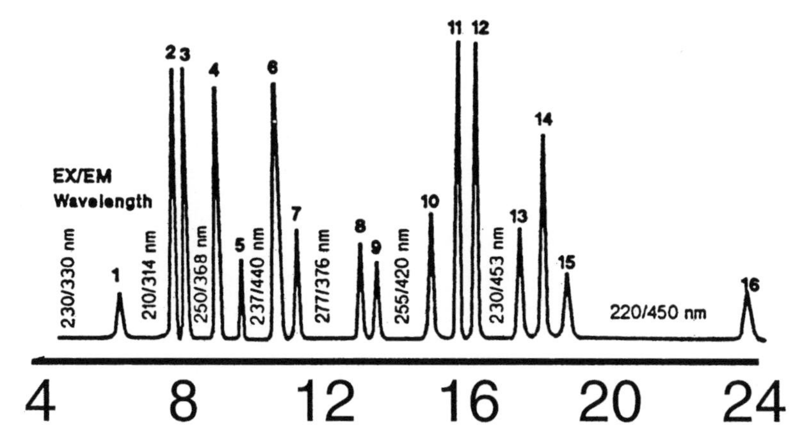


Figure 5.15. Reversed-phase gradient elution separation of a mixture of polycyclic aromatic hydrocarbons using time programmed fluorescence detection. Compounds: 1 = naphthalene; 2 = acenaphthene; 3 = fluorene; 4 = phenanthrene; 5 = anthracene; 6 = fluoranthene; 7 = pyrene; 8 = benz(a)anthracene; 9 = chrysene; 10 = benzo(b)fluoranthene; 11 = benzo(k)fluoranthene; 12 = benz(a)pyrene; 13 = dibenz(a,h)anthracene; 14 = benzo(g,h,i)perylene; 15 = indeno(1,2,3-cd)pyrene; and 16 = coronene.

For dilute solutions (< 0.05 AU) the observed fluorescence intensity is related to sample concentration by the expression $I_f = I_o \varphi_f(2.3\varepsilon lc)$, where $I_f$ is the fluorescence emission signal, $I_o$ the source excitation intensity, $\varphi_f$ the quantum yield (number of photons emitted/number of photons absorbed), ε the molar absorption coefficient, l the cell path length, and c the sample concentration. For typical conditions the detector response is linear over about two to three orders of magnitude. Sensitivity depends on the instrument ($I_o$ and the reduction of scattered and stray light), the sample (quantum efficiency), and the composition of the mobile phase (solvents, impurities, etc). The major sources of detector noise are specular scatter, Rayleigh and Raman scattering occurring in the flow cell material and mobile phase and luminescence impurities present in the flow cell materials and the mobile phase. Specular scatter results from the reflections and refractions of the excitation wavelength at the various optical interfaces of the detector.

Both the position of the emission wavelength envelope and the emission intensity can be affected by the mobile phase composition and even by the presence of contaminants, Table 5.3. Fluorescence detection is compatible with gradient elution unless one or more components of the mobile phase contain a high level of fluorescent impurities.

Flow cells for fluorescence detection with 3-10 μl volumes are similar in design to those used for absorption measurements. For microcolumn systems, the light output of conventional sources cannot be fully exploited because only a small fraction of the available excitation energy can be directed into the flow cell. If lasers are used as the source of excitation energy, a reduction in the cell volume can be achieved while maintaining an acceptable signal level [110,134]. Laser sources provide high intensity levels of collimated monochromatic light easily adapted to illuminating very small

Table 5.3
The effect of the mobile phase on fluorescence

| Mobile phase | Fluorescence emission |
|---|---|
| pH | Both the emission wavelength and fluorescence intensity of ionizable compounds (i.e. those containing acidic or basic functional groups) are critically dependent upon pH and solvent hydrogen-bonding interactions. |
| Solvent | Intensity changes of an order of magnitude and large wavelength shifts occur for compounds which undergo strong solvent interactions. A shift in the fluorescence spectrum to longer wavelengths is usually observed with an increase in the solvent dielectric constant. If the solvent absorbs any of the excitation or emission energy, the sensitivity will be reduced. |
| Temperature | Many compounds show marked temperature dependence with increasing temperature causing a decrease in intensity of 1-2% per °C. |
| Concentration | At high concentrations the emission signal becomes non-linear due to self-absorption by the sample itself or complete absorption of the excitation energy before it can fill the full cell volume. High emission intensity may overload the photosensor resulting in an irregular and fluctuating signal. |
| Quenching | Impurities in the mobile phase, particularly oxygen, may quench the signal from low concentrations of fluorescent compounds (see solvent degassing). |
| Photodecom-position | High intensity sources may cause sample decomposition, which depends on the residence time of the sample in the detector cell. |

volumes. When sample sizes are limited, for example, in some biological applications, laser-induced fluorescence may be the only technique suitable for detection. Most lasers provide only a limited number of discrete wavelengths, most of which are above 300 nm, and may not coincide with desirable excitation wavelengths for natural fluorescent compounds. Generally speaking, lasers provide a less stable light output than conventional sources and flicker noise limits analyte detectability. For conventional diameter columns and normal flow cells there is usually no advantage to using laser excitation sources over conventional arc sources.

Flow cells for packed capillary column systems are usually of two types. A flow cell can be fabricated from an unpacked portion of the fused-silica capillary by removing a section of the polyimide coating similar to those used for absorption measurements, except that the emission signal is generally measured at right angles to the direction of the excitation beam. In-column fluorescence detection is also possible using packed fused-silica capillary columns by removing a small section of the polyimide coating to create a viewing area before the column end frit [135,136]. The sample zones are now detected in the presence of the stationary phase eliminating band broadening from connecting tubes between the column and detector flow cell. In addition, the sample zones are focused, such that their concentration is increased by the factor $(1 + k)$, where k is the solute retention factor, compared to the concentration of the eluted zone. There is also an environmental factor, which could result in an enhancement or diminution of the fluorescence emission for the adsorbed sample. On the other hand, it is likely that the excitation beam does not completely penetrate the column packing, so that only a fraction of the adsorbed analyte is excited and the presence of the packing causes additional scattering of the excitation beam manifested as an increase in background noise. In most cases, sample detectability does not seem to be very different for in-column detection compared with postcolumn detection, which is generally more convenient. Since light scattering and fluorescence impurities in the fused-silica tubing contribute to detector noise flow cells without walls have also been used [110]. Detection volumes in the low nanoliter range can be achieved with a sheath-flow cell, for example. In this approach the column eluent is ensheathed by a strong flow of solvent with a similar refractive index to the eluent, compressing it into a narrow stream a few μm in diameter.

#### *5.7.1.3 Chemiluminescence Detectors*

Chemiluminescence detection differs from fluorescence detection in that an exothermic chemical reaction is used to either transfer energy to a fluorescent compound or create a fluorescent product in an excited state, which later relaxes to the ground state releasing at least part of its excess energy as light [104,112,137,138]. Chemiluminescence detection is used to improve the sensitivity or selectivity of fluorescence detection. Since chemiluminescence does not require a light source detector noise from stray light and source instability are reduced providing for an increase in sample detectability of one to three orders of magnitude for the most favorable cases. But signal enhancement is not obtained for all reactions and the number of useful applications is restricted compared with natural fluorescence. For liquid chromatography it is convenient to consider detector options based on gas phase or liquid phase chemiluminescence reactions.

Gas phase chemiluminescence detectors are used for the selective detection of nitrogen-containing compounds [138-140]. Most nitrogen-containing compounds (except $N_2$) are converted to nitric oxide by reaction with oxygen at about 900-1100°C. Subsequent reaction of the nitric oxide with ozone at a reduced pressure produces electronically excited nitrogen dioxide, which relaxes to the ground state by emission of near-infrared radiation centered around 1200 nm. The column eluent is nebulized into a high temperature pyrolysis chamber supplied with a flow of oxygen. Complete combustion of the mobile phase and sample components in this chamber are essential. The pyrolysis gases consisting mainly of unconsumed oxygen, water, carbon dioxide and nitric oxide (from the oxidation of nitrogen-containing compounds) are routed through a membrane drying system to remove all water and then to the reaction chamber for quantification. The reaction chamber consists of a light-tight gas chamber where ozone and the pyrolysis gases mix producing a chemiluminescence emission when the pyrolysis gases contain nitric oxide. The emission is detected by a photomultiplier after passage through an optical filter to enhance the selectivity for the desired chemiluminescence signal. The detection limit for nitrogen-containing compounds is about $10^{-12}$ g

(N/s), selectivity $10^7$ (g N/g C) with a linear response range of $10^4 - 10^5$. The equimolar response of the detector for nitrogen eliminates the need for authentic reference standards for all mixture components [141]. The detector is not well characterized at present and some questions concerning its ruggedness have been raised. There is less information available for the sulfur chemiluminescence detector (section 3.10.3.2) which has been used with packed capillary columns [142].

Liquid phase chemiluminescence detectors usually consist of a postcolumn reactor (section 5.8) connected to a fluorescence detector with its source disabled [104,137,138,143-145]. The column eluent is combined with one or several reagents that initiates the desired chemiluminescence reaction. The intensity of light emission depends on the rate of the chemical reaction, the efficiency of production of the excited state, and the efficiency of light emission from the excited state. The chemiluminescence intensity is sensitive to environmental factors such as temperature, pH, ionic strength, and solution composition. In addition, the detection system has to be designed to accommodate the time dependence of the chemiluminescence signal to ensure that adequate and representative emission occurs in the detector flow cell.

One of the most common reactions exploited for chemiluminescence is based on the hydrogen peroxide oxidation of an aryl oxalate ester, which produces a high-energy intermediate (1,2-dioxethane-3,4-dione) as the reactive species. Interaction of the 1,2-dioxetane-3,4-dione with a fluorescent compound results in its decomposition with energy transfer to the fluorescent compound. The excited fluorescent compound then emits light as it relaxes to the ground state. This reaction can be used to determine naturally fluorescent compounds and non-fluorescent compounds after formation of a fluorescent derivative, as well as some non-fluorescent compounds that are easily oxidized in the quench mode. Another common reaction is based on the quenching of the chemiluminescent reaction between luminol, an oxidant such as hydrogen peroxide, and a catalyst in an alkaline medium [146]. Under these conditions luminol is converted to 3-aminophthalate in an excited state that emits light at 425-435 nm. Because the chemiluminescence intensity is directly proportional to the concentration of luminol, peroxide and catalyst, species that can be converted into peroxides, species labeled with luminol or the catalyst, or species that interfere in the reaction mechanism can be quantified by monitoring the chemiluminescence emission.

### 5.7.2 Refractive Index

Refractive index detectors are bulk property detectors with a near universal response, albeit limited by poor sample detectability for some applications [101-106]. Differential detection is employed to minimize background noise with the result that the detector response is in some way related to the difference in refractive indices of the mobile phase and the mobile phase containing the sample. Consequently, an analyte will be detected only if its refractive index is different from that of the mobile phase. Peaks in both a positive and negative direction may be observed in the same chromatogram depending on whether the analyte has a higher or lower refractive index than that of the mobile phase. Detector response factors are generally different for each component in a separation unless they have identical refractive index values. The sensitivity of the detector to variations in flow and temperature limit its use to isocratic separations.

The ultimate performance limit of the refractive index detector is caused by fluctuations in the refractive index of the mobile phase. The detector is sensitive to small changes in temperature and mobile phase composition, and the majority of problems associated with its practical operation can be traced to this cause. Changes in composition can occur from incompletely mixed mobile phases, leaching of prior samples or solvents from the column, or changes in the amount of dissolved gases in the solvent. The temperature and pressure dependence of the refractive index results in an offset in the detector baseline with changes in flow rate. Incomplete equilibrium of the incoming mobile phase temperature with the temperature of the detector flow cell and viscous heating in the inlet to the cell both contribute to this flow sensitivity. To be able to detect $10^{-6}$ to $10^{-8}$ g of solute, a noise equivalent concentration of $10^{-8}$ refractive index units is required. To maintain this noise level the temperature of the detection system must be thermostatted to within 0.001°C.

Commercially available refractive index detectors usually employ refraction, reflection or interference of a collimated light beam to sense differences in refractive index in sample and reference flow cells. The comparatively poor response and sensitivity to environmental factors restricts most applications to analytical separations using conventional diameter columns and preparative separations. Small flow cells suitable for use with packed capillary and small bore columns are easily fabricated and detectors based on reflection [147] and backscattering interferometry [148], for example, have been described for use with these columns. Poor concentration sensitivity and complicated detector operation, however, limit these detector systems. Other methods of refractive index difference measurements include surface plasmon resonance-based detectors [149] and spectroscopic refractometry [150]. The spectroscopic refractive index detector provides a novel method of correcting for thermal noise by measuring and comparing the refractive index response at two or more wavelengths simultaneously. This approach shows potential for improving detection limits in analytical refractive index detection.

The refraction-type refractive index detector (deflection-refractometer), Figure 5.16, measures the deflection of a collimated light beam when it crosses a dielectric interface separating two media of different refractive index at an angle of incidence other than zero (Snell's law). Light from a tungsten lamp passes through a beam mask, collimating lens, sample and reference cells; is reflected by a mirror; and passes back through the cells and lens, which focuses the light onto a position sensitive photosensor. A rotatable glass plate between the lens and photosensor is used to zero the detector signal when the sample and reference cells contain the same solvent. The sample and reference cells are triangular in shape and located in a single fused glass assembly. For analytical applications the cell volume is usually 10 μl. The incident light beam leaves the first cell and enters the second cell at 45°. If the refractive indices are the same in the two cells, the light beam leaving the first cell will be refracted the same amount but in the

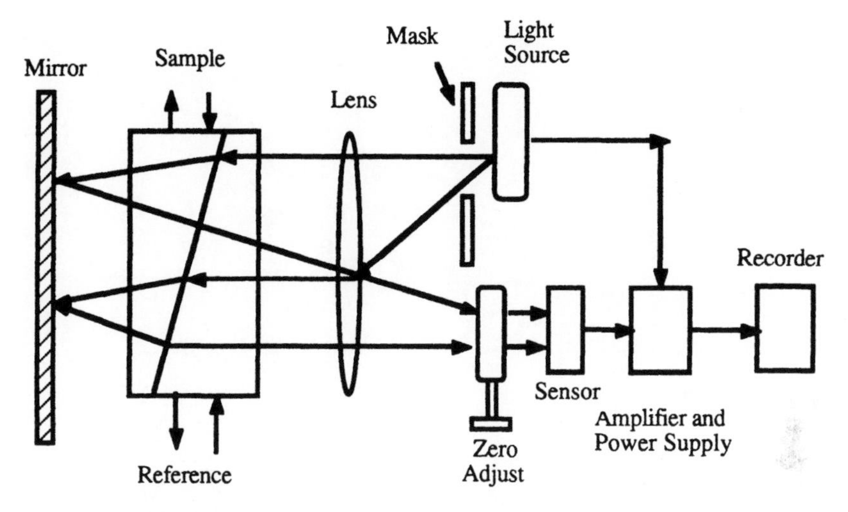


Figure 5.16. Schematic diagram of a refractive index detector employing the principle of refraction (deflection-refractometer).

opposite direction to that at which it enters the second cell. The net deflection of the light beam, therefore, is zero after passage through the two cells. If the refractive indices for the two cells are different, there will be a net deflection of the light beam after passage through the two cells, which is doubled after reflection from the mirror and passage back through the cells. This translates into a lateral displacement of the focused beam at the position of the photosensor. Deflection-refractometers have the advantage of a wide linear range, the entire refractive index range can be accommodated using a single cell, low volume cells can be fabricated, and the cells are relatively insensitive to air bubbles or the buildup of contaminants on the sample cell windows.

Reflection-type refractive index detectors, Figure 5.17, sense a change in the light intensity transmitted through a dielectric interface between the surface of a glass prism and the liquid to be monitored when the refractive index of the liquid changes. A beam mask and collimating lens are used to produce two parallel beams of light from a tungsten lamp that impinge on the prism-cell interface at slightly less than the critical angle for total internal reflection. The thin sample and reference cells are defined by the surface of the prism, the grained surface of the metal cell plate, which contains the inlet and outlet ports, and a thin PTFE gasket. A collecting lens focuses the light scattered from the grained metal surface on to a photosensor. The difference in intensity of the light reflected in the two beams is related to the refractive index difference between the sample and reference cell by Fresnel's law and has a fairly large linear range when the incident light strikes the cell near its critical angle. A disadvantage of this approach is that two different prisms are required to cover the full refractive index range of common chromatographic solvents. Advantages include high sensitivity, a small cell volume, and ease of cleaning.

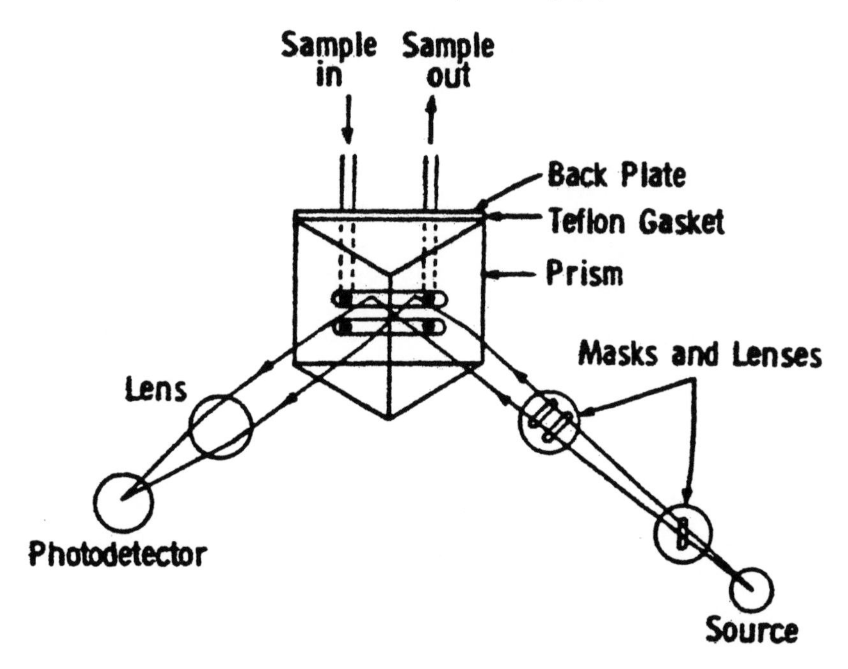


Figure 5.17. Schematic diagram of a refractive index detector employing the principle of reflection (Fresnel-refractometer).

The interferometric refractive index detector, Figure 5.18, measures the difference in the effective path length (speed of light) of a light beam passing through sample and reference cells by means of the interference of the two beams when recombined at a photosensor. By definition the speed of light is directly proportional to the refractive index of a medium. As the relative speed of light in the two cells changes, destructive interference occurs and the light intensity at the photosensor decreases. To maximize sensitivity, the light beam is first linearly polarized, and the horizontal and vertical polarized components used for the sample and reference light beams, respectively. It is likely that the sensitivity of this detector is similar to that of the reflection and refraction based detector when used with conventional diameter columns.

### 5.7.3 Evaporative Light Scattering

The evaporative light-scattering detector (ELSD) is a near universal detector suitable for the determination of (mainly) neutral compounds that are less volatile than the mobile phase used for the separation [151,152]. Primary uses include the detection of compounds with a weak response to the UV detector, especially carbohydrates, lipids, surfactants, polymers and petroleum products. Its greater sensitivity and ease of use in gradient elution separations makes it preferable to the refractive index detector for these applications. The ELSD is compatible with most volatile solvents used for normal and

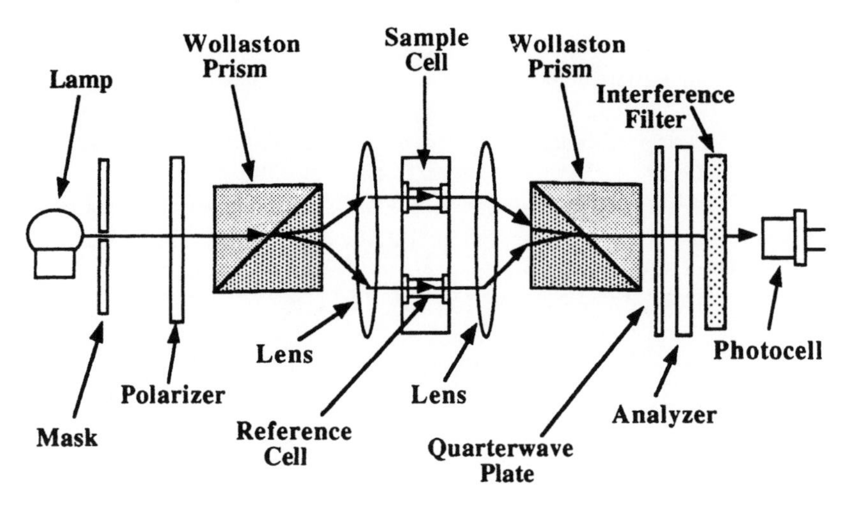


Figure 5.18. Schematic diagram of an interferometric refractive index detector.

reversed-phase separations. Typical limits of detection are about 1 ng corresponding to a sample concentration of about 1 μg / ml for conventional diameter columns and about 10 μg / ml for packed capillary columns. Its main drawback for quantitative analysis is the non-linear relationship between the detector response and the sample mass [153-157]. At low sample concentrations the detector response is described by the relationship $A = am^b$ where A is the peak area, m the analyte mass and a and b are coefficients that depend on the detector operating conditions and analyte characteristics. The b coefficient generally takes values between 0.9 and 1.8. Even though the response is non-linear it will usually have some range in which the response appears linear. The linearity of the detector response can be extended by logarithmic transform of the response function. Since the coefficients a and b are compound and system dependent calibration is required for quantification.

The three main processes that occur successively with the ELSD are: conversion of the column eluent to an aerosol (nebulization); evaporation of volatile components from the aerosol droplets in a heated drift tube to form particles (desolvation); and determination of the particle density by light scattering in a gas flow cell. Pneumatic nebulizers are commonly used for aerosol generation. The column eluent is mixed with a gas stream in a concentric tubular nebulizer forming a high-velocity jet of tiny liquid droplets. These droplets either proceed directly into the drift tube or a separate spray chamber connected to the drift tube where the larger droplets are removed by condensation. The use of a spray chamber generates a more homogeneous aerosol and allows the drift tube to be operated at a lower temperature for desolvation. The low flow rates typical of packed capillary columns require replacement of the standard nebulizer by a miniaturized version [158-162]. For example, by a fused-silica restrictor (e.g. a length of 20 μm internal diameter capillary tube) housed in a wider bore stainless steel capillary with nitrogen gas flowing through the space between the two capillaries. Insertion of the miniaturized nebulizer into the end of the drift tube and its contact with the heated wall of the drift tube, the purpose of which is to preheat the nebulizer gas, allows faster system equilibration at startup and production of a more stable and homogeneous aerosol.

The aerosol is next desolvated by evaporation of volatile mobile phase components by contact with warm gas as it traverses the short thermostatted drift tube. The temperature required to dry the droplets depends on the evaporation properties of the mobile phase and the droplet size. Typical temperatures are between 30 to 80°C. It is important to use as low a temperature as possible for this process to maintain particle size uniformity and to encourage formation of larger crystalline particles, which maximize the intensity of scattered light. In addition, too high a temperature may result in the complete or partial loss of the more volatile analytes from the chromatogram. At the end of the drift tube, a light beam is scattered by the particles present in the gas flow. The intensity of the scattered light is measured by a photosensor located at an angle to the incident beam direction. The intensity of the scattered light depends on the size distribution of the particles formed in the drift tube, which in turn depends on the size of the droplets formed during the nebulization process. The addition of a chamber for condensation nucleation between the drift tube and light scattering cell, which has the effect of increasing the average particle size, has been shown to provide a significant increase in detector response [151,163,164]. Mobile phases that contain non-volatile components cause an elevated background and decrease sample detectability as well as a rapid degradation of detector performance, due to their deposition in the optical cell. Mobile phase additives and buffers are limited to such species as volatile acids (formic, acetic, trifluoroacetic and nitric acids), volatile bases (e.g. triethylamine, ammonia, pyridine) and volatile salts (ammonium bicarbonate, ammonium acetate, ammonium formate).

### 5.7.4 Electrochemical

Electrochemical detectors based on the principles of capacitance (dielectric constant detector), resistance (conductivity detector), voltage (potentiometric detector) and current (amperometric, coulometric and polarographic detectors) have all been used in liquid chromatography [165-171]. Conductance is a universal property of ions in solution. It is an obvious choice for the continuous and selective monitoring of ionic species in an aqueous mobile phase. The prominence of the conductivity detector is tied to the acceptance of ion chromatography for the analysis of common ions with poor UV absorption properties. Amperometric detection is based on the oxidation or reduction of an analyte at a working electrode held at a potential sufficient to initiate the electrode reaction. The electric current resulting from the reaction is proportional to the concentration of the analyte in the column eluent. Some selectivity in the detection process is possible based on the structure-related voltages required for the electrolysis of individual functional groups and ions. Electrolysis is usually incomplete, about 1-10%, for typical detection conditions. By increasing the electrode surface area the electrolysis reaction may reach (near) completion, corresponding to the coulometric limit. The response of the coulometric detector is absolute, eliminating the need for calibration. On the other hand, although the conversion of electroactive species is higher, the background noise is also greater, and sample detectability is generally similar for coulometric and amperometric detectors. The potential applied to the working electrode is usually constant throughout the separation or applied as a pulse waveform. Triple-pulse and other waveforms are often used when products of the electrochemical reaction contaminate the electrode surface. The successive and repetitive application of a measuring potential, a cleaning potential and a conditioning potential at a noble metal electrode is the basis of pulsed amperometric detection. This is the most important of the pulsed electrochemical detection techniques and is used to extend the application range of amperometric detection to polar aliphatic compounds (e.g. carbohydrates, surfactants, etc). Polarographic detectors, employing a dropping mercury electrode, are used less frequently than solid-electrode amperometric detectors in liquid chromatography. Their sensitivity is similar to solid-electrode amperometric detectors but their application range is restricted to reducing species (the low oxidation potential of mercury precludes its use for oxidizable species). Potentiometric detectors (ion selective electrodes) are too specific for general applications in liquid chromatography [167,172].

*5.7.4.1 Conductivity Detectors*

Conductivity detectors have a simple and robust design and are easily miniaturized [102,106,168,172-175]. A typical detector consists of two electrodes housed in a chamber fabricated from a non-conducting material with a volume of 1 – 10 μl. The electrodes are of low surface area and constructed of an inert conducting material, such as stainless steel or platinum. The cell resistance is usually measured in a Wheatstone bridge circuit. During the measurement of the cell resistance, undesirable processes, such as electrolysis or the formation of an electric double layer at the electrodes, may occur. By varying the frequency of the applied potential these undesirable effects are suppressed or completely eliminated. Some detectors apply a sinusoidal wave potential across the cell electrodes at 1-10 KHz and synchronous detection of the component of the cell current which is in phase with the applied potential frequency. Alternatively, a bipolar pulse technique consisting of two successive short voltage pulses of opposite polarity but equal amplitude and duration can be used. At exactly the end of the second pulse the cell current is measured and the cell resistance is determined from Ohm's law. The detector cell is thermostatted to avoid sudden temperature changes in addition to a temperature-compensating circuit to adjust the cell resistance to the temperature of the column eluent. All detectors have some means of correcting for the background conductivity of the mobile phase, which limits the absolute sample detectability. In the absence of significant background conductivity detection limits of $10^{-8}$–$10^{-9}$ g / ml are possible.

If a conductivity detector is used to monitor the effluent from an ion-exchange column to separate anions, for example, then the change in observed conductivity, $\Delta G$ (expressed in microsiemens), as the analyte peak passes through the conductivity cell is given by

$$\Delta G = (\lambda_{A^-} - \lambda_{E^-}) C_A \alpha_A / 10^{-3} K \quad (5.3)$$

where $C_A$ is the concentration of analyte anions, $\alpha_A$ the fraction of analyte present in the ionic form, $\lambda_{E^-}$ and $\lambda_{A^-}$ the limiting equivalent ion conductance of the competing eluent anion and analyte, respectively, and K the cell constant that takes into account the physical dimensions of the cell [168,176,177]. Useful sample detectability requires a large difference in the limiting equivalent ionic conductance of the analyte and eluent ions. This difference can be positive or negative, depending on whether the eluent ion is strongly or weakly conducting. When the mobile phase ions have a higher equivalent conductance than the analyte ions the method of detection is referred to as indirect conductivity detection.

The most common use of conductivity detection is in single column and suppressed ion chromatography (section 4.3.7.2). Single column ion chromatography is achieved by using dilute solutions of electrolyte with a high affinity for the stationary phase and a suitable difference in equivalent conductivity to the analyte ions for adequate sample detectability as a mobile phase. Suppressed conductivity detection is based on the chemical amplification of the conductance of the analyte ions with suppression of the conductance of eluent ions prior to conductivity detection. For this purpose weak acid (or base) salts are used as eluent competing ions for the elution of anions (or cations) followed by exchange of the eluent ions using an ion-exchange resin or membrane for hydronium ions (hydroxide ions for cations). This process results in a reduction of the conductance of the eluent by replacing the high-conductance eluent ions by a partially ionized weak acid (or weak base). Simultaneously, the conductance of analyte ions, which are strong acids (or strong bases) is increased by pairing them with highly conducting hydronium ions (or hydroxide ions for cation determinations). For example, suppressed conductivity detection of chloride and nitrate uses eluents containing sodium hydroxide or carbonate-bicarbonate buffers that can be converted into species of low conductance ($H_2O$ and $H_2CO_3$ in this case) after exchanging the cations of the eluent (sodium) for hydronium ions by a suitable cation-exchange device. In the same process the analyte ions are converted to species of higher conductance by replacing sodium ions by hydronium ions ($HCl$ and $HNO_3$). Originally, a packed column containing a high capacity ion-exchange resin in the hydrogen or hydroxide form located in the eluent flow path between the separation column and the conductivity detector was used for suppressed conductivity detection. This arrangement, however, was not ideal and it is more common to use ion-exchange membrane suppressors today [168,173, 178-180].

The main disadvantages of packed-column suppressors are: that they require periodic regeneration due to limited capacity; result in significant extracolumn band broadening, since reasonable suppressor capacity requires relatively large suppressor columns; weak acids and bases exhibit variable retention resulting from several mechanisms, which change with the degree of suppressor exhaustion; and the "water dip" resulting from the elution of the sample solvent often hampers trace analysis of some analytes that are eluted early in the chromatogram. Some of these problems were solved by the introduction of hollow-fiber suppressors based on ion-exchange polymeric membranes [168,181]. The main advantage of the hollow-fiber ion-exchange suppressor was that it allowed continuous operation of the ion chromatograph by exchanging ions across the membrane wall into a continuously regenerated solution of electrolyte bathing the fibers. Relatively large dead volumes, limited suppression capacity, and the fragile nature of the fibers made them less than ideal. Eluent suppression capacity was largely limited by slow mass transfer of ions to the membrane wall, and although second generation hollow-fiber suppressors filled with beads or fibers and shaped into various helical forms to promote turbulence improved ion transport to the membrane wall, they lack equivalent performance to the micromembrane suppressors now in common use.

The micromembrane suppressor, Figure 5.19, combines the high ion-exchange capacity of packed-bed suppressors with the constant regeneration feature of hollow-fiber suppressors in a low-dead volume ($< 50$ μl) configuration [168,173,178-182]. The eluent from the separator column flows between two thin ion-exchange membranes that are separated by an intermediate screen made from a polymer with ion-exchange sites. A solution of electrolyte is pumped countercurrently and externally to the membrane in small volume channels that are partially filled with polymer screens. The purpose of the screens is to enhance the transport of ions to the ion-exchange membrane by generating turbulent flow and by providing a site-to-site path for transport of ions to the membranes (screens in the eluent chamber contain ion-exchange sites). The mechanism of suppressor operation for anion and cation detection is illustrated in Figure 5.20. The micromembrane suppressor has a higher suppression capacity than other suppressor designs. It is compatible with the use of higher buffer concentrations, higher eluent flow rates and facilitates applications requiring composition gradients for convenient separation. For weak acids and bases the micromembrane suppressor can be used to exchange metal cations to enhance conductivity by formation of conducting salts [183,184]. In addition, it can be used as a device for pH control in a reaction detector configuration [182].

Significant advances in suppressor technology with respect to convenience of operation have been made in the last few years using electrolytic suppressors and solid-phase chemical regeneration [178,179,185-188]. A disadvantage of membrane suppressors is that they require an additional pump and a supply of regenerant solution for continuous operation. Electrolytic suppressors replace the regenerant solution with water that is electrolysed in the regenerant compartments of the micromembrane suppressor or separate regeneration cells to produce the hydronium and hydroxide ions necessary for the suppression reaction. In self-regenerating suppressors the deionized eluent waste from the suppressor is recycled from the detector to the regenerant chamber. The amount of water available for sweeping out products from the suppressor

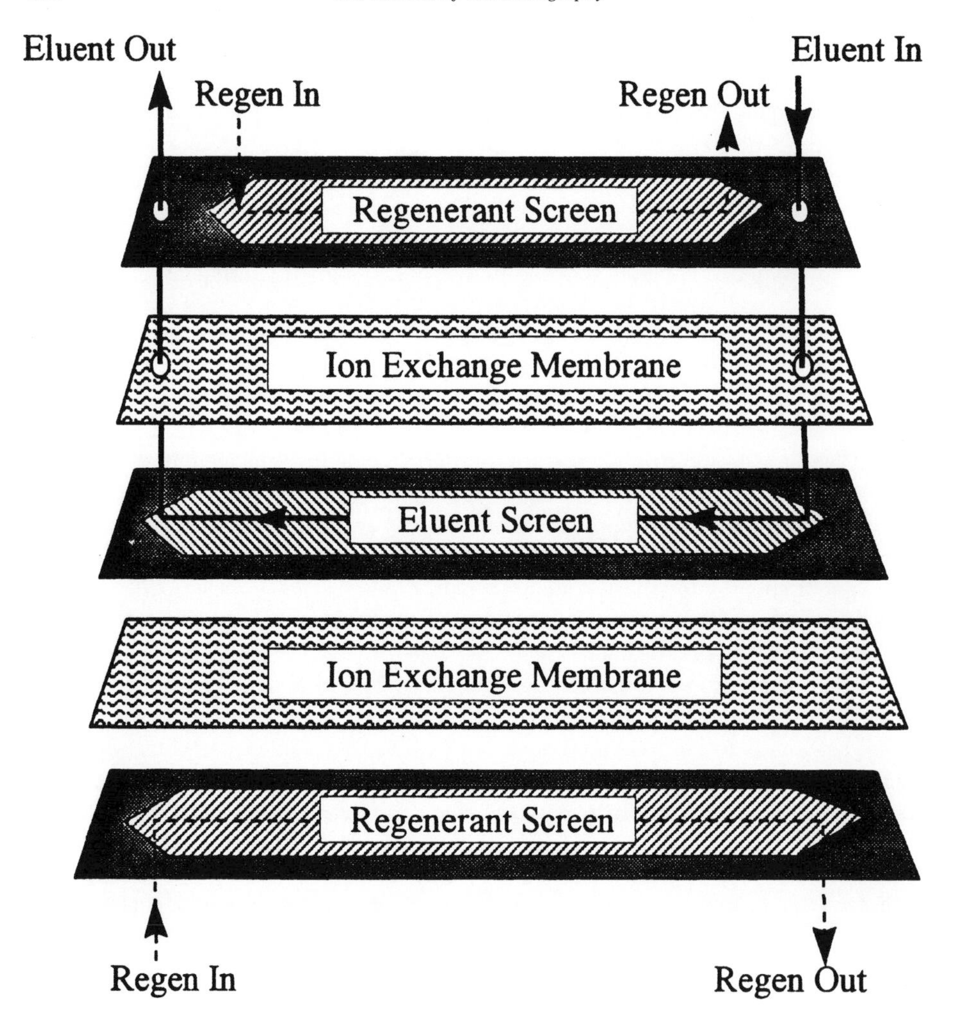


Figure 5.19. Exploded view of a micromembrane suppressor (regen = regenerant solution). (From ref. [182]; ©Elsevier).

operation is now limited to the eluent flow rate, resulting in a higher, although acceptable background noise for most applications, when compared to systems using an external source of water. Self-regenerating suppressors can be used with solid-phase regenerant cartridges in the recycle line to strip suppressor waste products from the regenerant supply. The cartridges provide a visual indication of exhaustion and can be used without electrolytic suppression. Organic solvents are not well tolerated by electrolytic suppressors and chemical regenerant methods are preferred for applications employing partially aqueous mobile phases.

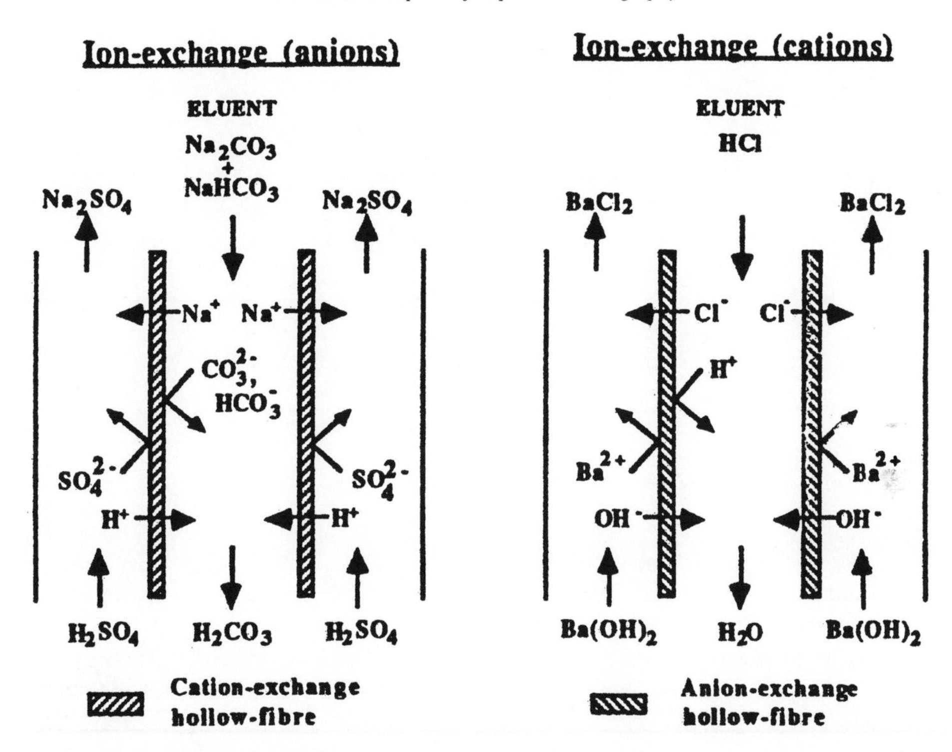


Figure 5.20. Schematic representation of the operation of a membrane suppressor for suppressed conductivity detection of anions and cations. (From ref. [168]; ©Elsevier).

*5.7.4.2 Constant Potential Amperometric Detectors*

The constant potential amperometric detector determines the current generated by the oxidation or reduction of electoactive species at a constant potential in an electrochemical cell. Reactions occur at an electrode surface and proceed by electron transfer to or from the electrode surface. The majority of electroactive compounds exhibit some degree of aromaticity or conjugation with most practical applications involving oxidation reactions. Electronic resonance in aromatic compounds functions to stabilize free radical intermediate products of anodic oxidations, and as a consequence, the activation barrier for electrochemical reaction is lowered significantly. Typical applications are the detection of phenols (e.g. antioxidants, opiates, catechols, estrogens, quinones) aromatic amines (e.g. aminophenols, neuroactive alkaloids [quinine, cocaine, morphine], neurotransmitters [epinephrine, acetylcoline]), thiols and disulfides, amino acids and peptides, nitroaromatics and pharmaceutical compounds [170,171]. Detection limits are usually in the nanomolar to micromolar range or 0.25 to 25 ng / ml.

A number of electrochemical cell designs have been described but the most popular configurations are the three-electrode thin-layer cell and the wall-jet cell, Figure 5.21[20,102,166-171,189]. The column eluent is introduced either parallel to

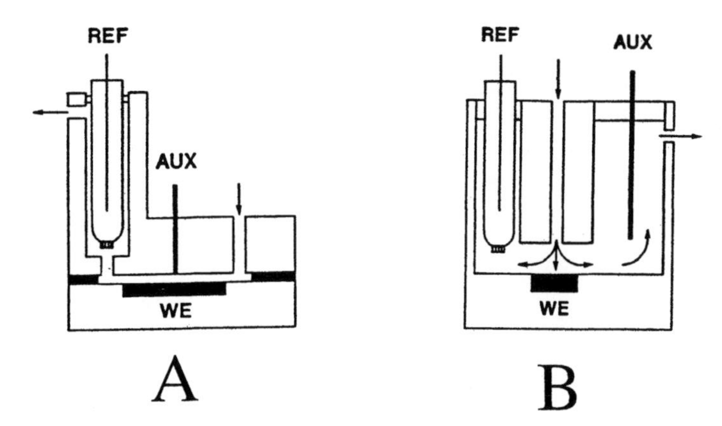


Figure 5.21. Schematic representation of a thin-layer electrochemical cell (A) and a wall-jet electrochemical cell (B). AUX = auxiliary electrode, REF = reference electrode and WE = working electrode.

the working electrode embedded in the channel wall using the thin-layer cell, or perpendicular to the working electrode surface followed by radial dispersion in the wall-jet design. The cell body is usually made out of an insulating material like PTFE and machined to accommodate the eluent channel and various electrodes. The working electrode, usually in the form of a disk, is embedded in the wall of a channel defined by a gasket sandwiched between two blocks. A typical cell volume is about 1-10 μl. The reference and auxiliary electrodes are placed at the downstream side of the working electrode, so that leakage from the reference electrode or formation of electrolysis products at the auxiliary electrode does not interfere with the working electrode. Inclusion of a reference electrode allows the potential of the working electrode to be set at a predetermined value suitable for electrolysis of the analytes. The chosen potential is selected to operate the cell in the steady state limiting current region, and can be deduced from hydrodynamic voltammograms. The auxiliary electrode serves as a reference for current measurement at the working electrode. As part of a current feedback loop the auxiliary electrode can inject or remove charge from the cell to maintain a constant potential at the working electrode. The types of materials used for the reference electrode (e.g. silver-silver chloride) and auxiliary electrode (e.g. platinum or stainless steel) are not critical, but the choice of working electrode material is very important as it affects detector performance. A wide variety of materials have been used for the working electrode, but the most popular is glassy carbon. Glassy carbon has a surface that is easily polished, inert to all solvents used in liquid chromatography and with a working range of -1.3 to +1.5 V, it is suitable for a wide variety of applications. Metal electrodes are generally too easily oxidized producing high residual currents for general use in detector flow cells. The large negative potential range of the silver/mercury amalgam electrode makes it suitable for reduction reactions. Porous electrode cells because of the high surface area of the working electrode are popular for coulometric detection. The working electrode is usually made from a composite of carbon particles sufficiently permeable to allow the column eluent to flow through the electrode. These cells are difficult to clean and have a high background noise limiting their use for amperometry.

Amperometric detectors are easily miniaturized with preservation of performance, since their operation is based on reactions at the electrode surface. Using a single carbon fiber or microelectrode as a working electrode allows detector cells of very small volume and in-column detectors to be constructed for use in open tubular and packed capillary column liquid chromatography [189-192]. These microcolumn separation techniques combined with amperometric detection are exploited for the quantitative analysis of volume-limited samples such as the contents of single cells [193,194].

The chromatogram is recorded by measuring the detector cell current at a fixed potential as the sample is eluted from the column. The background will remain constant as long as the mobile phase velocity and composition do not change and is subtracted from the analytical signal. The resulting detector current is directly proportional to the concentration of electroactive species in accordance with Faraday's law. Detector operation is critically dependent on flow rate constancy, solution pH, ionic strength, temperature, cell geometry, the condition of the electrode surface and the presence of electroactive impurities (dissolved oxygen, halides, trace metals, etc.) [20,169-171]. The background detector noise and, thus the ultimate sensitivity of the detector, are controlled by dissolved oxygen, ionic impurities and the contamination of the electrode surface, coupled with transient changes in the eluent flow rate. Gradient elution is not normally possible. Amperometric detection requires the use of mobile phases containing salts or mixtures of water with water-miscible organic solvents, conditions that are compatible with reversed-phase and ion-exchange chromatography, but are more difficult to achieve with other separation modes. These problems can be circumvented, to some extent, if the mobile phase is water miscible by adding a makeup flow of support electrolyte at the column exit.

Amperometric detectors containing (usually) two or more working electrodes have been developed to improve the overall selectivity or sensitivity of the detection process [20,102,189]. The two working electrodes can be arranged in parallel or series. The peak height ratio at two different potentials for electrodes in parallel provides information on the identity of a peak or an estimate of its purity. Using the series configuration compounds eluted from the column can be oxidized (or reduced) on the first electrode and subsequently reduced (or oxidized) on the second. Since the electroactive compounds that irreversibly react on the first electrode are eliminated, determination of a reversible electroactive compound is possible with high selectivity. This technique can be useful for the removal of oxygen interference in which the upstream electrode reduces oxygen and the analyte, and the downstream electrode oxidatively detects only the reduction product of the analyte. Lithographic techniques allow the construction of microelectrode arrays containing a large number of adjacent micron-size electrodes on a single substrate [195,196]. One application of these arrays is enhancement of sample detectability by redox cycling across adjacent oxidizing and reducing electrodes.

*5.7.4.3 Pulsed Amperometric Detectors*

Pulsed amperometric detection is used for the direct detection of a variety of polar aliphatic compounds, many of which, like carbohydrates, peptides and sulfur-containing compounds are of biological interest [171,197-200]. Most aliphatic compounds are not amenable to constant potential amperometric detection. Free-radical products from the oxidation of aromatic molecules can be stabilized by π-resonance; hence the activation barrier for reaction is decreased. This mechanism is unavailable for stabilizing aliphatic free radicals. The activation barrier for oxidation of aliphatic compounds can be decreased at noble-metal electrodes with partially unsaturated d-orbitals (e.g. gold, platinum) that can adsorb and thereby stabilize free radical oxidation products and intermediates. Carbon electrodes are not electrocatalytic and are unsuitable for pulsed amperometric detection.

Pulsed amperometric detection exploits the high electrocatalytic activity of noble-metal electrodes by combining amperometric detection with pulsed potential cleaning of the electrode surface. Surface-adsorbed products produced during the short detection potential step are efficiently desorbed from the noble-metal electrodes by application of a large positive-potential pulse with concurrent formation of a surface oxide layer. The oxide-covered electrode is inert and must be reduced by a negative-potential pulse to desorb the oxide layer and restore the activity of the noble-metal surface. By utilizing a simple three-step potential waveform with a frequency of 0.5-2 Hz, Figure 5.22, a reproducible response with favorable detection limits is obtained during chromatographic separations. By monitoring the transient current at the same delay time in the detection cycle a Faraidic signal with a reasonable linear dependence on the analyte concentration is obtained. Selection of potential values for the pulse waveform is identified from cyclic voltammetry. Mobile phase characteristics (e.g. pH, ionic strength and type and concentration of organic modifier) affect the detector response characteristics. Organic solvents can be tolerated only if they are not electroactive for the conditions of the pulsed-potential waveform. Electroinactive additives adsorbed by the electrode can affect baseline stability if they interfere in the formation of the oxide layer. The pH of the eluent is important for controlling the amplitude of the detector response. Increasing pH shifts the potential for oxide formation to more negative potentials from the optimized value for detection. Separations employing a pH gradient are possible by replacing the reference electrode by a pH electrode to continuously adjust the detection potential throughout the separation. All alcohol-based compounds (e.g. carbohydrates) are determined by direct oxidation at oxide-free surfaces with a potential less than about 200 mV at gold electrodes in alkaline solutions or platinum electrodes in acidic solutions. Aliphatic amines and amino acids are detected at gold and platinum electrodes in both alkaline and acidic solutions at potentials greater than 150 mV at which concurrent formation of surface oxide promotes electrode reactions. Numerous sulfur-containing compounds can be detected under similar conditions. Sulfur-containing compounds and inorganic ions can also be detected under conditions selected to interfere in the formation of the surface-oxide layer in an indirect detection mode.

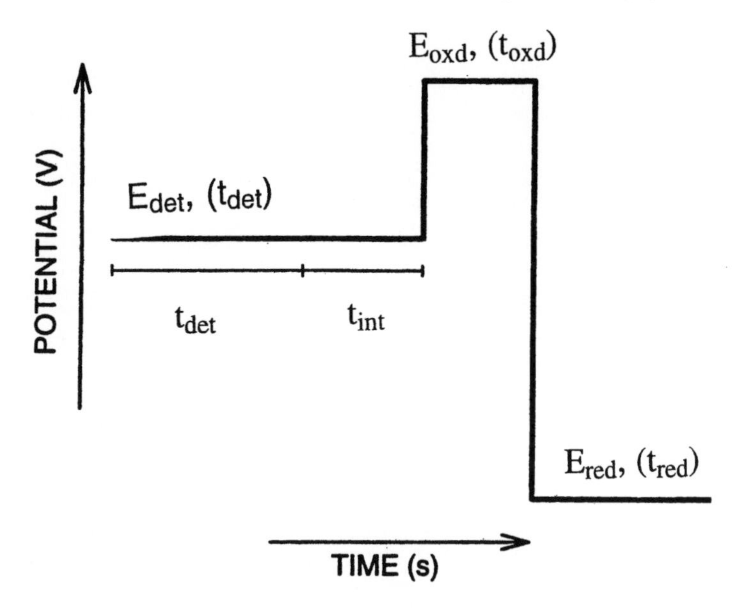


Figure 5.22. Three-step potential-time waveform for pulsed amperometric detection at a noble-metal electrode. $E_{det}$ is the appropriate potential for the desired surface-catalyzed reaction applied for a time $t_{det}$ composed of a delay time, $t_{del}$, and a short period at the end of the detection period, $t_{int}$, during which the current is sampled. Following the detection process, the electrode is cleaned by oxidative desorption concurrent with surface oxide formation using a step change in the applied potential to $E_{oxd}$ for time $t_{oxd}$. The oxide-coated electrode is reactivated by a negative step potential $E_{red}$ for time $t_{red}$ sufficient to remove the oxide layer prior to the next cycle of the waveform.

Limits of detection for pulsed amperometric detection vary widely depending on the analyte and operating conditions. For normal-sized injections, limits of detection for compounds with a favorable detector response are usually in the pM to nM range ($\mu$g / ml to ng / ml sample concentrations) with a linear response range of about 1000 fold.

### 5.7.5 Miscellaneous Detectors

From the perspective of system requirements detection has long been considered the weak link in liquid chromatography, especially when compared with the ability of gas chromatographic detectors to function across the range of normal application needs for gas chromatography. This has maintained interest in the development of new detection principles for liquid chromatography, some of which have resulted in commercial devices for specific applications. In addition, mass spectrometry is emerging as a useful detection technique for liquid chromatography as well as a source of structural information for identification (section 9.2). Viscometry and low-angle laser light scattering detectors for absolute molecular weight estimations in size-exclusion chromatography are discussed in section (4.3.9.3).

Given the success of ionization detectors in the field of gas chromatography it is no surprise that their adaptation to liquid chromatography has attracted considerable interest, although arguably without great success [201]. The two common approaches to interface liquid chromatography to gas phase ionization detection are based on transport systems and direct eluent introduction as a droplet stream (eluent-jet interface) or aerosol (nebulizer interface). In general ionization detectors are incompatible with typical eluent flow rates and solvents commonly used in liquid chromatography. Transport detectors attempt to solve this problem by eliminating the mobile phase by assisted evaporation prior to detection. Direct introduction methods rely on the tolerance of some ionization detectors to vapors generated from the low flow rates (μl / min) characteristic of packed capillary columns.

A transport detector consists of a carrier (e.g. wire, disk or chain) that continuously passes through a series of separate cleaning, coating, evaporation and pyrolysis chambers [202,203]. In the cleaning chamber all volatile material is removed from the carrier by heating it to about 750°C before it enters the coating chamber. In the coating chamber a jet applies a fraction of the eluent as a thin film to the carrier. In the evaporation chamber a controlled temperature and counter flow of gas strips the volatile mobile phase from the less volatile sample. In the pyrolysis chamber a controlled temperature is used to volatilize the sample into the detector, or into a chemical reactor for conversion to an easily detectable product. Modern versions of the transport detector use an oxidized titanium ribbon 1.5 mm wide and 0.1 mm thick as a carrier, which has a higher coating capacity than earlier versions employing stainless steel wires or metal chains. At a velocity of 3.5 mm/s the coating capacity of the ribbon is about 0.18 ml / min. Operation is restricted to separations obtained with volatile mobile phases. It can be configured to operate with a number of ionization detectors based on sample volatilization or decomposition occurring in the pyrolysis chamber. The rather complex operating mechanism, noisy signal and moderate sensitivity are considered the main drawback of transport detectors.

Packed capillary columns can be interfaced directly to flame-based ionization and photometric detectors or flameless thermionic ionization and electron-capture detectors using an eluent-jet or nebulization liquid introduction system [201,204,205]. The eluent-jet interface generates a jet of micron-sized droplets by application of a sharp temperature gradient at the tip of the introduction capillary. The capillary is cooled by a coaxial flow of gas while the eluent is heated inductively by an external radio frequency generator. The nebulization interface mixes the eluent stream with gas and sprays the liquid directly into the detector or a separate chamber that allows the larger droplets to settle out creating a more uniform aerosol. Both organic and aqueous eluents can be used with element-selective detectors. Sample detectability and linear response ranges are reasonable if not as good as when the same detectors are used in gas chromatography (e.g. TID < 10 ng / s N, FPD 20 pg / s S, ECD 5-10 pg all with a linear response range of about $10^3$). The limited injection volumes of packed capillary columns results in less impressive concentration detection limits. The flame ionization detector can not be used with mobile phases containing a significant amount of organic solvent but is a useful detector for separations employing hot pressurized water as a mobile phase (section 7.7.2).

Studies on chemical speciation in the environment and biology require a separation step combined with element selective detection. Many inorganic and organometallic compounds are easily separated by reversed-phase, ion-pair, ion-exchange and size-exclusion chromatography but the selectivity of common detectors is often inadequate for typical practical applications. In these studies, element-selective detection has generally been achieved by coupling liquid chromatography with flame, furnace and plasma atomic emission and atomic absorption spectrometers [206-210]. Contemporary interest is centered on the use of inductively coupled plasma-mass spectrometry (ICP-MS) as a universal element and isotope detector [210-213]. The ICP is a well-characterized, high temperature source suitable for the atomization and ionization of elemental species. The plasma is an electrodless atmospheric pressure discharge created in a stream of (usually) argon sustained by energy from a radio frequency generator. Advantages of ICP-MS as a chromatographic detector include a wide linear range, low limits of detection (ng / l to μg / l), multielement and isotope analysis capability, and simple ion spectra. The mass spectrometer may be set to monitor the isotopic signal of an element, or several elements, with respect to time resulting in a chromatogram that contains peaks only for the elements of interest. The coupling of liquid chromatography to ICP-MS is straightforward and only requires a short length of inert polymeric or stainless steel tubing to connect the column to the nebulizer of the ICP. The choice of nebulizer is dictated by the mobile phase flow rate. Liquid flow rates of about 1 ml / min are compatible with conventional cross flow and concentric pneumatic nebulizers with single pass, double pass and cyclone-type spray chambers [214,215]. Conventional nebulizers provide low sample utilization with only about 1-3 % of the sample entering the nebulizer reaching the plasma. Small though significant improvements in sample utilization have been achieved using cooled spray chambers, hydraulic high-pressure nebulization, ultrasonic nebulizers and thermospray introduction systems. For liquid flow rates of 10-120 μl / min direct injection nebulization without a spray chamber is used. Both aqueous and organic solvents and their mixtures can be handled fairly easily by an appropriate combination of liquid flow rate, spray chamber temperature and radio frequency power supplied to the plasma. High proportions of organic solvents can lead to deposition of elemental carbon at the ion entrance to the mass spectrometer resulting in unstable and irreproducible ICP-MS operation. The addition of oxygen to the nebulizer gas and an increase in radio frequency power to the plasma can be used to minimize this problem. The type of organic solvent used for the separation is usually selected based on its ability to provide an adequate separation and by its effect on the ICP [216]. Mobile phases with a high salt content (> 2% w/v) should be avoided. In general, polyatomic spectral interferences are the most serious problem in ICP-MS (e.g. $^{40}Ar^{35}Cl^{+}$ on the determination of $^{75}As^{+}$, $Ar_2H^{+}$ on the determination of $^{81}Br^{+}$, etc). Polyatomic spectral interferences mainly occur in the mass region below 100 where many non-metallic elements of interest for the selective detection of organic compounds occur. A collision reaction cell containing a mixture of helium and hydrogen can be used to eliminate interference from argon-containing ions for reliable detection [217].

The use of flow-through radioactivity detectors to study the metabolism or degradation of mostly $^{14}C$, $^{3}H$, $^{35}S$ and $^{32}P$ labeled compounds *in vivo* or environmental systems is widely used in some industries and university laboratories. Complexes containing the $\gamma$-emitters $^{125}I$ and $^{99m}Tc$ are used in nuclear medicine. All these radionuclides can be determined by the secondary photons produced by the interaction of energetic particles with a suitable photon emitter (scintillator). The column eluent is usually either mixed with a suitable scintillation cocktail prior to the detector cell or is passed through a transparent tube packed with a suitable solid-phase scintillator housed in the detector cell [106,218-220]. Two photomultiplier tubes at opposite sides of the detector cell are used for photon counting. The output of the photomultiplier tubes is processed using coincidence electronics. In this way noise spikes and other electronic interferences sensed at both photomultipliers simultaneously are rejected. The output from the coincidence electronics passes to a multichannel pulse height analyzer and then to a computer for data collection and display. Operational characteristics are usually indicated as disintegrations per minute (dpm) with typical detection limits of 300-500 dpm for $^{14}C$ and 1000-1500 dpm for $^{3}H$. Limits of detection are related to the counting time (the residence time of the sample in the detector flow cell). For increased sample detectability it may be preferable to collect the column eluent in a fraction collector and count fractions off-line using a conventional scintillation counter [220,221]. The handling of radioactive materials requires special facilities and dedicated equipment, which may become contaminated by samples, and is rarely attempted in general analytical laboratories.

Chiral-selective detection based on polarimetry or circular dichroism provides an alternative to enantiomeric-selective separation systems for identifying and quantifying individual enantiomers in racemic mixtures [222-227]. Polarimetry and circular dichroism detectors respond directly to the intrinsic optical activity of a compound with a response to each enantiomer that is equal in magnitude, but opposite in sign. These differences are based upon the differential response of a chiral compound to polarized light. In polarimetry the difference in refractive index for right versus left circularly polarized light is measured at a single wavelength generated by a laser to improve sensitivity and facilitate the use of small detector volumes. When passed through an optically active medium (e.g. sample in a flow cell) one circularly polarized component will exhibit retardation relative to the other. The change in the phase between the two recombined components manifests itself as a rotation in the plane of polarization specified by an angle. In circular dichroism the differential absorbance of a chiral system for right and left circularly polarized light is recorded over a range of wavelengths (220-500 nm) isolated by a scanning monochromator from an arc lamp, which is then plane-polarized. The differential absorption is determined directly by sequentially probing the sample with right and left circularly plane-polarized light. Sample detectability in circular dichroism is background limited since the signal represents a small change (the polarization dependent differential absorption) on top of a large background (the intrinsic absorption of the compound). For both polarimetry and circular dichroism detection sample concentration requirements are usually in the μM range with the most favorable mass detection limits in a range above 10 ng. It is possible to determine an enantiomer excess above about 0.1 %.

## 5.8 POSTCOLUMN REACTION SYSTEMS

Postcolumn reaction systems are convenient for on-line derivatization used to improve sample detectability or detection selectivity for target analytes, often present in low concentrations or complex matrices [228-230]. The derivatization reaction does not have to yield a single, stable product, provided that the reaction is reproducible. Reactions must be reasonably fast (< 20 min), however, for convenient real time detection and to maintain the separation integrity, and excess reagent must not interfere with detection. Efficient mixing of the derivatizing reagent and column eluent is required and usually reaction solvents must be miscible with the mobile phase. A significant source of detector baseline noise, leading to an increase in detection limits, arises from the additional pumps and auxiliary equipment required to facilitate continuous operation of the reaction system. Dual wavelength detection with one wavelength set close to the isosbestic point for the reagent and the other close to the absorbance maximum for the analytes allows a significant reduction of pump noise for absorbance detection [231]. Although many suitable reaction chemistries have been described [228,232], the general complexity of postcolumn reaction systems and the need for lengthy optimization restricts their use for occasional sample problems. The main applications are for the detection of amino acids and biogenic amines [233], carbohydrates [234,235], pesticides and herbicides [236-238] and metal ions [168,179] in clinical, environmental and food samples.

The postcolumn reaction system must provide for the continuous addition of controlled volumes of one or more reagents to the column eluent, followed by mixing of the eluent-reagent mixture and incubation for some time and temperature governed by the needs of the reaction. Detection of the reaction products is usually by absorption, fluorescence or electrochemical techniques. Some typical postcolumn reaction configurations are shown in Figure 5.23. A piston or pneumatic pump is used to deliver the reagent to the reactor at a constant low-pulse flow rate to facilitate reproducible reaction conditions and a stable detector baseline. Peristaltic pumps are suitable for use with segmented flow systems that generally operate at low pressures. Nearly all reaction detectors contain a mixing device, T- or Y-piece, or cyclone, for contacting and homogeneously mixing the column eluent and reagent streams [228,239]. The mixing device should have a small volume to minimize band broadening and to allow low reagent flow rates to reduce peak dilution. Cyclone mixers are superior in this respect as they can be constructed with a smaller dead volume ($\approx$ 0.1 μl) and work efficiently with reagent flow rates as low as 4% of the column flow rate. The reaction vessel is usually an open tube or packed bed of appropriate dimensions to store the combined eluent and reagent volumes for the time required to obtain optimum

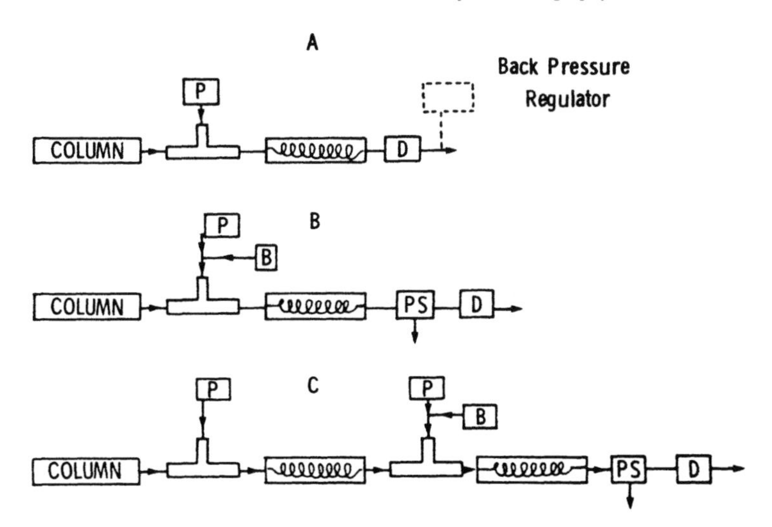


Figure 5.23. Schematic diagram of some typical postcolumn reaction configurations for liquid chromatography. A, non-segmented tubular reactor; B, segmented tubular reactor; C, extraction segmented reaction detector. P = pump, PS = phase separator, B = device for introducing bubbles and D = detector.

reaction yields. If the segmentation principle is employed, a device for introducing a bubble of liquid or gas into the eluent stream is required [228,240,241]. This results in the separation or segmentation of the column eluent into a series of reaction compartments whose volume is governed by the dimensions of the transfer tube and the frequency of bubble introduction. Prior to detection a phase separator is needed to remove the segmentation agent. Two different principles are used for phase separation. A simple T-piece with one upward opening will remove the gas from the liquid flow using density differences. In general the same principle is used for the construction of phase separators for liquid segmented systems. Here the different wettability of glass and organic polymeric tubing is used to separate aqueous and organic segments. Alternatively, the differential permeability of membranes for organic liquids and water provides a suitable mechanism for phase separation. A phase separator is also required when ion-pair formation or extraction is employed in the reaction system. Dispersion in the phase separator is often a significant source of band broadening. For reactions using solid-phase reagents, immobilized enzymes, etc., a packed bed tubular reactor is used [228,242,243]. Reactions that are too slow at room temperature are accelerated by thermostatting the reaction vessel to a higher temperature. A cooling coil may be required before the detector to prevent interference in the detector operation or to avoid bubble formation.

The parabolic flow profile in an open-tubular reactor restricts applications to fast reactions if extensive band broadening is to be avoided [244]. The reaction time can be extended by using reactors prepared from optimally deformed capillary tubes [245-247]. The production of secondary flow, even at low flow rates, breaks up the parabolic profile and minimizes band broadening. The easiest way to induce secondary flow is by using knitted or stitched open-tubular reactors. Knitted open tubular (KOT) reactors are generally fabricated from PTFE capillary tubes deformed in such a way as to produce alternating right and left loops with small coiling diameters. Each subsequent loop is bent out of the plane of its nearest neighbor. Stitched open tubular (SOT) reactors are usually prepared from narrow-bore stainless-steel capillary tubes woven through a steel mesh in serpentine fashion with alternate loops displaced to the right and left. The continuous change in the coiling direction and the bend of the coil out of plane leads to a continuous change in the direction of the secondary flow producing a plug like flow profile with almost identical velocities at the wall and tube center. The pressure drop across a KOT reactor is about twice that of an open tube, which is the limiting factor in establishing the available reaction time. For PTFE capillaries with a pressure limit of about 10 atmospheres reaction times approaching 10 minutes are possible for flow rates typical of conventional packed columns.

The time required for the reaction most profoundly influences the reactor design. For fast reactions (i.e. $< 1$ min) open tubular or KOT reactors are commonly used. They simply consist of a mixing device and a coiled stainless steel or PTFE capillary tube enclosed in a thermostat. The length and internal diameter of the capillary tube and the combined column and reagent flow rate through the reactor control the reaction time. Reagents such as fluorescamine and o-phthalaldehyde are frequently used in this type of system to determine primary amines, amino acids and indoles in biological and environmental samples. A common method for determining N-methylcarbamates in environmental and food extracts employs a two-stage postcolumn reaction system [236,237]. In the first reactor N-methylcarbamates are rapidly hydrolyzed to methylamine which is then converted in the second reactor to a highly fluorescent isoindole by reaction with o-phthalaldehyde and 2-mercaptoethanol.

For reactions of intermediate kinetics (i.e. reaction times $< 5$ and $> 1$ min) a KOT or packed bed reactor is used. The packed bed reactor is constructed from a length of column tubing packed with an inert material of small diameter, such as glass beads The pressure drop across the packed bed reactor limits the length and smallest particle size that can be used. The inner diameter of the reactor (e.g. 2.5-6.0 mm) can be varied to adjust the reactor volume to the required reaction time. The packed bed reactor is particularly useful for reactions involving solid-phase reagents such as catalysts, immobilized enzymes and metallic reducing agents [228,229,242,243]. Since no reagent solutions are added in this case, there are no additional pumps or mixing units required and no dilution or mixing problems associated with postcolumn reagent addition.

For slow reactions (reaction times $> 5$ min) an air or liquid segmented reactor is typically used. Segmented systems are also required for reactions that require separation of the reagent from the reaction product by solvent extraction, for example when ion-pair reagents are used to form the product for detection. The reaction products are usually of different polarity to the reagent, permitting their separation by extraction in the continuous flow mode by segmenting with immiscible organic solvent plugs (the extraction solvent) or by air segmentation with an additional flow of extraction solvent.

### 5.8.1 Photoreactors

Photolysis may be used to convert some analytes into easier to detect products or products suitable for subsequent on-line chemical derivatization [137,248]. Photoreactors are simple in design consisting of one or more high-powered discharge lamps in a reflective housing with fans or a Peltier cooler for heat removal. A typical arrangement uses a cylindrical tube mercury or xenon discharge lamp around which is wrapped a PTFE knitted open-tubular (KOT) capillary reactor. PTFE has excellent transparency to UV light and is preferable to quartz for fabrication of the reaction coil. For the low flow rates typical of small-bore columns a flow through reaction cell made from a length of fused-silica capillary tube with a window created by removing a section of the protective polyimide coating is generally used [249]. A selectable-power photoreactor was described that contains a central lamp around which is wrapped the reaction coil with four other lamps arranged at perpendicular axes to the reaction coil in a reflective housing [250]. By switching on various lamp combinations the light intensity can be optimized for the reaction while minimizing undesirable photodecomposition or other secondary reactions resulting in a lower yield of the detection product. The parameters that affect the extent of reaction include the residence time in the reaction coil (column flow rate and coil dimensions), mobile phase composition (solvents, pH, oxygen concentration, buffer salts) and the irradiation wavelength and its intensity [237,248,251]. Photolysis can result in the formation of products with a different spectra to the original analyte. Thus, either full spectra or individual wavelength absorption ratios for lamp on and lamp off conditions were suggested for confirmation of analyte identity [252,253]. The high specificity of the photochemical reaction detector is useful for particular applications but the number of compounds that yield useful detection products is limited [248].

## 5.9 INDIRECT DETECTION

Indirect detection is an alternative to derivatization for the detection of analytes with a weak detector response. It is commonly used in ion exchange (particularly ion chromatography) and ion-pair chromatography with absorbance, fluorescence or amperometric detection [168,254,255]. This requires the selection of an eluent ion with favorable detection properties to regulate the separation process and provide a constant detector signal. Detector transparent analyte ions cause displacement of eluent ions from the eluted band and a decrease in the detector response compared with the steady state signal for the mobile phase. The detected ion concentration is coupled to the retention mechanism, which can result in the appearance of additional system peaks in the chromatogram (section 4.3.3.2). These applications should be distinguished from vacancy chromatography, in which the detector response is simply a difference chromatogram without involving a displacement mechanism [256]. Indirect detection techniques also include postcolumn response enhancement or diminution by inclusion complexation, quenching or hydrophobic adsorption of (usually) a detectable fluorescent additive and ion replacement or enzyme reactions to generate detectable species [257-259]. These techniques are not further discussed here except to note that vacancy chromatography generally provides poor sample detectability and the other methods are restricted to a limited number of analytes with specific properties.

Indirect detection provides a universal detection method for analytes sharing the same retention mechanism. The concentration limit of detection is given by $C_{LOD} = C_E / (TR \times DR)$ where $C_E$ is the concentration of the detectable ion in the eluent. The other terms that influence the detector response are the transfer ratio (ratio of the number of molecules of the mobile phase additive displaced by each analyte molecule), TR, and the dynamic reserve (the signal-to-noise ratio of the background signal), DR. The relatively high absorption of the eluent ion in indirect UV detection causes an increase in the detector baseline noise. This noise limits analyte detectability. The UV-transparent analytes at low concentration provide a small change in the high background signal, which must be distinguished from baseline noise. For a reasonable linear response range and favorable sample detectability the concentration of UV absorbing eluent ion should be low enough that Beer's law is still valid, while at the same time the concentration must be appropriate for the sample separation conditions. All other things being equal, the concentration of the absorbing eluent ion should be as small as possible and the transfer ratio and dynamic reserve as large as possible. All three properties are interrelated and an optimum detector response may require simultaneous optimization of the chromatographic system. In general, detection limits for indirect UV detection of typical ions separated by ion chromatography are comparable to conductivity detection but not better than conductivity detection.

## 5.10 REFERENCES


[1] V. Berry, CRC Crit. Revs. Anal. Chem. 21 (1989) 115.
[2] J. W. Dolan and L. R. Snyder, *Troubleshooting LC Systems*, Humana Press, Clifton, NJ, 1989.
[3] E. Katz, R. Eksteen, P. Schoenmakers and N. Miller (Eds.), *Handbook of HPLC*, Dekker, New York, 1998.
[4] P. Sadek, *Troubleshooting HPLC Systems*, Wiley, New York, 1999.
[5] I. D. Wilson, E. R. Adlard, M. Cooke and C. F. Poole (Eds.), *Encyclopedia of Separation Science*, Academic Press, London, 2000.
[6] J. Abian, A. J. Oosterkamp and E. Gelpi, J. Mass Spectrom. 34 (1999) 244.
[7] J. P. C. Vissers, J. Chromatogr. A 856 (1999) 117.
[8] H. D. Melring, E. van der Heeft, G. J. ten Hove and A. P. J. M. de Jong, J. Sep. Sci. 25 (2002) 557.
[9] J. P. Chervet, M. Urseum and J. B. Salzmann, Anal. Chem. 68 (1996) 1507.
[10] J. E. MacNair, K. C. Lewis and J. W. Jorgenson, Anal. Chem. 69 (1997) 983.
[11] J. E. MacNair, K. D. Patel and J. W. Jorgenson, Anal. Chem. 71 (1999) 700.
[12] N. J. Wu, J. A. Lippert and M. L. Lee, J. Chromatogr. A 911 (2001) 1.

[13] D. C. Collins, Y. Xiang and M. L. Lee, Chromatographia 55 (2002) 123.
[14] Y. Xiang, N. Wu, J. A. Lippert and M. L. Lee, Chromatographia 55 (2002) 399.
[15] M. J. Rehman, K. P. Evans, A. J. Handley and P. R. Massey, Chromatographia, 24 (1987) 492.
[16] K. Shoikhet and H. Engelhardt, Chromatographia 38 (1994) 421
[17] H. Engelhardt and C. Siffrin, Chromatographia 45 (1997) 35.
[18] M. Kele and G. Guiochon, J. Chromatogr. A 830 (1999) 41.
[19] V. J. Barwick, J. Chromatogr. A 849 (1999) 13.
[20] P. T. Kissinger, J. Chromatogr. 488 (1989) 31.
[21] J. J. Pedrotti, L. Angnes and I. G. R. Gutz, Anal. Chim. Acta 298 (1994) 393.
[22] W. A. MacCrehan, S. D. Yang, and B. A. Benner, Anal. Chem. 60 (1988) 284.
[23] W. A. MacCrehan and E. Schonberger, J. Chromatogr. B 670 (1995) 209.
[24] R. W. Slingsby, A. Bordunov and M. Grimes, J. Chromatogr. A 913 (2001) 159.
[25] A. Weston and P. R. Brown, *HPLC and CE. Principles and Practice*, Academic Press, San Diego, CA, 1997.
[26] C. Ericson and S. Hjerten, Anal. Chem. 70 (1998) 366.
[27] X. Zhou, N. Furushima, C. Terashima, H. Tanaka and M. Kurano, J. Chromatogr. A 913 (2001) 165.
[28] P. Jandera and J. Churacek, *Gradient Elution in Column Liquid Chromatography*, Elsevier, Amsterdam, 1985.
[29] J. W. Dolan, D. C. Lommen, and L. R. Snyder, J. Chromatogr. 485 (1989) 91.
[30] L. R. Snyder, J. J. Kirkland and J. L. Glajch, *Practical HPLC Method Development*, Wiley, New York, 1997.
[31] M. C. Harvey and S. D. Stearns, J. Chromatogr. Sci. 21 (1983) 473.
[32] S. R. Bakalyar, C. Phipps, B. Spruce and K. Olsen, J. Chromatogr. A 762 (1997) 167.
[33] M. D. Foster, M. A. Arnold, J. A. Nichols and S. R. Bakalyar, J. Chromatogr. A 869 (2000) 231.
[34] J. P. C. Vissers, A. H. de Ru, M. Ursem and J. P. Chervet, J. Chromatogr. A 746 (1996) 1.
[35] M. J. Mills, J. Maltas and W. J. Lough, J. Chromatogr. A 759 (1997) 1.
[36] A. Cappiello, G. Famiglini and A. Berloni, J. Chromatogr. A 768 (1997) 215.
[37] S. Heron, A. Tchapla and J. P. Chervet, Chromatographia 51 (2000) 495.
[38] J. Layne, T. Farcas, I. Rustamov and F. Ahmed, J. Chromatogr. A 913 (2001) 233.
[39] K. G. Kehl and V. R. Meyer, Anal. Chem. 73 (2001) 131.
[40] P. Campins-Falco, R. Herraez-Hernandez and A. Sevillano-Cabeza, J. Chromatogr. B 619 (1993) 177.
[41] K.-S. Boos and C.-H. Grimm, Trends Anal. Chem. 18 (1999) 175.
[42] L. Bovanova and E. Brandstetenova, J. Chromatogr. A 880 (2000) 149.
[43] I. Liska, J. Chromatogr. 655 (1993) 163.
[44] P. R. Haddad, F. Doble and M. Macka, J. Chromatogr. A 856 (1999) 145.
[45] M.-C. Hennion, J. Chromatogr. A 856 (1999) 3.
[46] T. Hyotylainen, N. Savola, P. Lehtonen and M. L. Riekkola, Analyst 126 (2001) 2124.
[47] K. Heinig and F. Bucheli, J. Chromatogr. B 769 (2002) 9.
[48] A. Asperger, J. Efer, T. Koal and W. Engewald, J. Chromatogr. A 960 (2002) 109.
[49] J. R. Veraart, H. Lingeman and U. A. Th. Brinkman, J. Chromatogr. A 856 (1999) 483.
[50] P. Simon, M. Lafontaine, P. Delsaut, Y. Morele and T. Nicot, J. Chromatogr. B 748 (2000) 337.
[51] J. Haginaka, Trends Anal. Chem. 10 (1991) 17.
[52] Z. Yu and D. Westerlund, Chromatographia 47 (1998) 299.
[53] E. Schoenzetter, V. Pichon, D. Thiebaut, A. Fernandez-Alba and M.-C. Hennion, J. Microcol. Sep. 12 (2000) 316.
[54] E. A. Hogendoorn, E. Dijkman, B. Baumann, C. Hidalgo, J. V. Sancho and F. Hernandez, Anal. Chem. 71 (1999) 1111.
[55] C. Schaler and D. Lubdo, J. Chromatogr. A 909 (2001) 73.
[56] L. E. Sojo and J. Djauhari, J. Chromatogr. A 840 (1999) 21.
[57] R. Wissiack, E. Rosenberg and M. Grasserbauer, J. Chromatogr. A 896 (2000) 159.
[58] J. Slobodnik, H. Lingeman and U. A. Th. Brinkman, Chromatographia 50 (1999) 141.

[59] N. Masque, R. M. Marce and F. Borrull, J. Chromatogr. A 793 (1998) 257.
[60] A. C. Hogenboom, M. P. Hofman, D. A. Jolly, W. M. A. Niessen and U. A. Th. Brinkman, J. Chromatogr. A 885 (2000) 377.
[61] M.-C. Hennion, C. Cau-Dit-Coumes and V. Pichon, J. Chromatogr. A 823 (1998) 147.
[62] C. F. Poole, A. D. Gunatilleka and R. Sethuraman, J. Chromatogr. A 885 (2000) 17.
[63] H. Lord and J. Pawliszyn, J. Chromatogr. A 885 (2000) 153.
[64] J. Wu and J. Pawliszyn, J. Chromatogr. A 909 (2001) 37.
[65] J. Wu, H. Kataoka, H. L. Lord and J. Pawliszyn, J. Microcol. Sep. 12 (2000) 255.
[66] G. D. Reed and C. R. Loscombe, Chromatographia 15 (1982) 15.
[67] M. Capparella, M. Foster, M. Larrousse, D. J. Phillips, A. Pomfret and Y. Tuvim, J. Chromatogr. A 691 (1995) 141.
[68] J. G. Atwood, G. L. Schmidt and W. Slavin, J. Chromatogr. 171 (1989) 109.
[69] J. J. Kirkland, M. A. van Straten and H. A. Claessens, J. Chromatogr. A 797 (1998) 111.
[70] R. G. Wolcott, J. W. Dolan, L. R. Snyder, S. R. Bakalyar, M. A. Arnold and J. A. Nichols, J. Chromatogr. A 869 (2000) 211.
[71] B. W. Yan, J. H. Zhao, J. S. Brown, J. Blackwell and P. W. Carr, Anal. Chem. 72 (2000) 1253.
[72] N. M. Djordjevic, P. W. J. Fowler and P. Houdiere, J. Microcol. Sep. 11 (1999) 403.
[73] P. Molander, R. Trones, K. Haugland and T. Greibrokk, Analyst 124 (1999) 1137.
[74] P. Molander, K. Haugland, E. Lundanes, S. Thorud, Y. Thomassen and T. Greibrokk, J. Chromatogr. A 892 (2000) 67.
[75] T. Andersen, P. Molander, R. Trones, D. R. Hegna and T. Greibrokk, J. Chromatogr. A 918 (2001) 221.
[76] R. W. Frei and K. Zech (Eds.), *Selective Sample Handling and Detection in HPLC. Part A*, Elsevier, Amsterdam, 1988.
[77] H. J. Cortes (Ed.), *Multidimensional Chromatography: Techniques and Applications*, Dekker, New York, NY, 1990.
[78] K. A. Ramsteiner, J. Chromatogr. 456 (1988) 3.
[79] K. Fried and I. W. Wainer, J. Chromatogr. B 689 (1997) 91.
[80] P. Campins-Falco, R. Herraez-Hernandez and A. Sevillano-Cabeza, J. Chromatogr. B 619 (1993) 177.
[81] U. A. Th. Brinkman, J. Chromatogr. A 665 (1994) 217.
[82] D. A. Wolters, M. P. Washburn and J. R. Yates, Anal. Chem. 73 (2001) 5683.
[83] D. W. Patrick, D. A. Strand and H. J. Cortes, J. Sep. Sci. 25 (2002) 519.
[84] N. Sagliano, S. H. Hsu, T. R. Floyd, T. V. Raglione and R. A. Hartwick, J. Chromatogr. Sci. 23 (1985) 238.
[85] F. Regnier and G. Huang, J. Chromatogr. A 750 (1996) 3.
[86] Z. Y. Liu and M. L. Lee, J. Microcol. Sep. 12 (2000) 241.
[87] R. E. Murphy, M. R. Schure and J. P. Foley, Anal. Chem. 70 (1998) 1585.
[88] M. R. Schure, Anal. Chem. 71 (1999) 1645
[89] J. P. C. Vissers, R. E. J. van Soest, J. P. Chervet and C. A. Cramers, J. Microcol. Sep. 11 (1999) 277.
[90] A. P. Kohne and T. Welsch, J. Chromatogr. A 845 (1999) 463.
[91] L. A. Holland and J. W. Jorgenson, Anal. Chem. 67 (1995) 3275.
[92] L. A. Holland and J. W. Jorgenson, J. Microcol. Sep. 12 (2000) 371.
[93] G. J. Opiteck, K. C. Lewis, J. W. Jorgenson and R. J. Anderegg, Anal. Chem. 69 (1997) 1518.
[94] G. J. Opiteck, J. W. Jorgenson and R. J. Anderegg, Anal. Chem. 69 (1997) 2283.
[95] K. Wagner, K. Racaityte, K. K. Unger, T. Miliotis, L. E. Edholm, R. Bischoff and G. Marko-Varga, J. Chromatogr. A 893 (2000) 293.
[96] K. Wagner, T. Miliotis, G. Marko-Varga, R. Bischoff and K. K. Unger, Anal. Chem. 74 (2002) 809.
[97] A. V. Lemmo and J. W. Jorgenson, Anal. Chem. 65 (1993) 1576.
[98] J. P. Larmann, A. V. Lemmo, A. W. Moore and J. W. Jorgenson, Electrophoresis 14 (1993) 439.
[99] A. W. Moore and J. W. Jorgenson, Anal. Chem. 67 (1995) 3456.
[100] T. F. Hooker and J. W. Jorgenson, Anal. Chem. 69 (1997) 4134.
[101] T. M. Vickrey (Ed.), *Liquid Chromatography Detectors*, Dekker, New York, NY, 1983.

[102] P. C. White, Analyst, 109 (1984) 677 and 973.
[103] E. S. Yeung (Ed.), *Detectors for Liquid Chromatography*, Wiley, New York, NY, 1986.
[104] G. Patonay (Ed.), *HPLC Detection. Newer Methods*, VCH, New York, NY, 1992.
[105] D. Parriott (Ed.), *A Practical Guide to HPLC Detection*, Academic Press, San Diego, CA, 1993.
[106] R. P. W. Scott, *Chromatographic Detectors. Design, Function and Operation*, Dekker, New York, NY, 1996.
[107] L. Huber and S. A. George, *Diode Array Detection in HPLC*, Dekker, New York, NY, 1993.
[108] M. B. Smalley and L. B. McGowan, Adv. Chromatogr. 37 (1997) 29.
[109] A. A. Abbas and D. C. Shelly, J. Chromatogr. A 691 (1995) 37.
[110] R. J. Vandenesse, N. H. Velthorst, U. A. T. Brinkman and C. Gooijer, J. Chromatogr. A 704 (1995) 1.
[111] N. Simeon, R. Myers, C. Bayle, M. Nertz, J. K. Stewart and F. Couderc, J. Chromatogr. A 913 (2001) 253.
[112] A. R. Bowie, M. G. Sanders and P. J. Worsfold, J. Biolum. Chemilum. 11 (1996) 61.
[113] K. P. Jones, Trends Anal. Chem., 9 (1990) 195.
[114] C. B. Boring and P. K. Dasgupta, Anal. Chim. Acta 342 (1997) 123.
[115] F. Cuesta Sanchez and D. L. Massart, Anal. Chim. Acta 298 (1994) 331.
[116] M. A. Sharaf, Adv. Chromatogr. 37 (1997) 1.
[117] K. De Braekeleer, A. de Juan and D. L. Massart, J. Chromatogr. A 832 (1999) 67.
[118] A. Garrido Frenich, J. R. Torres-Lopasio, K. De Braekeleer, D. L. Massart, J. L. Martinez Vidal and M. Martinez Galera, J. Chromatogr. A 855 (1999) 487.
[119] J. G. D. Marr, G. C. R. Seaton, B. J. Clark, and A. F. Fell, J. Chromatogr. 506 (1990) 289.
[120] J. N. Little and G. J. Fallick, J. Chromatogr. 112 (1975) 389.
[121] J. P. Chervet, R. E. J. Vansoest and M. Ursem, J. Chromatogr. 543 (1991) 439.
[122] C. T. Culbertson and J. W. Jorgenson, Anal. Chem. 70 (1998) 2629.
[123] H. L. Wang, E. C. Yi, C. A. Ibarra and M. Hackett, Analyst 125 (2000) 1061.
[124] M. Aiello and R. McLaren, Anal. Chem. 73 (2001) 1387.
[125] C. E. Evans and V. L. McGuffin, J. Chromatogr. 503 (1990) 127.
[126] D. O. Hancock, C. N. Renn, and R. E. Synovec, Anal. Chem. 62 (1990) 2441.
[127] M. Born and E. Wolf, *Principles of Optics*, Cambridge University Press, Cambridge, UK, 1999.
[128] G. Openhaim and E. Grushka, J. Chromatogr. A 942 (2002) 63.
[129] S. J. Setford and S. Saini, J. Chromatogr. A 867 (2000) 93.
[130] R. Ferrer, J. Guiteras and J. L. Beltran, J. Chromatogr. A 779 (1997) 123.
[131] P. J. Marriott, P. D. Carpenter, P. H. Brady, M. J. McCormick, A. J. Griffiths, T. S. G. Hatvani and S. G. Rasdell, J. Liq. Chromatogr. 16 (1993) 3229.
[132] A. D. Wheatley and S. Sadhra, J. Liq. Chromatogr. & Rel. Technol. 21 (1998) 2509.
[133] J. A. Hernandez-Arteseros, J. L. Beltran, R. Compano and M. D. Prat, J. Chromatogr. A 942 (2002) 275.
[134] G. Gooijer and A. J. G. Mank, Anal. Chim. Acta 400 (1999) 281.
[135] H. Rebscher and U. Pyell, J. Chromatogr. A 737 (1996) 171.
[136] M. Verzele and C. Dewaele, J. Chromatogr. 395 (1987) 85.
[137] J. W. Birks (Ed.), *Chemiluminescence and Photochemical Reaction Detectors in Chromatography*, VCH, New York, NY, 1989.
[138] A. M. Garcia-Campana and W. R. G. Baeyens (Eds.), *Chemiluminescence in Analytical Chemistry*, Marcel Dekker, New York, NY, 2001.
[139] X. Yan, J. Chromatogr. A 842 (1999) 267.
[140] K. Petritis, C. Elfakir and M. Dreux, J. Chromatogr. A 961 (2002) 9.
[141] E. M. Fujinari, J. D. Manes and R. Bizanek, J. Chromatogr. A 743 (1996) 85.
[142] T. B. Ryerson, A. J. Dunham, R. M. Barkley and R. E. Sievers, Anal. Chem. 66 (1994) 2841.
[143] H. A. G. Niederlander, C. Gooijer and N. H. Velthorst, Anal. Chim. Acta 285 (1994) 143.
[144] A. Dapkevicius, T. A. van Beek and H. A. G. Niederlander, J. Chromatogr. A 912 (2001) 73.
[145] S. P. Forry and R. M. Wightman, Anal. Chem. 74 (2002) 528.

[146] M. Yamaguchi, H. Yoshida and H. Nohta, J. Chromatogr. A 950 (2002) 1.
[147] A. A. Abbas and D. C. Shelly, Anal. Chim. Acta 397 (1999) 191.
[148] C. K. Kenmore, S. R. Erskine and D. J. Bornhop, J. Chromatogr. A 762 (1997) 219.
[149] G. Cepria and J. R. Castillo, J. Chromatogr. A 759 (1997) 27.
[150] A. Hanning and J. Roeraade, Anal. Chem. 69 (1997) 1496.
[151] M. Kohler, W. Haerdi, P. Christen and J. L. Veuthey, Trends Anal. Chem. 16 (1997) 475.
[152] J. A. Koropchak, L.-E. Magnusson, M. Heybroek, S. Sadain, X. Yang and M. P. Anisimov, Adv. Chromatogr. 40 (2000) 275.
[153] P. Van der Meeren, J. Vandendeelen and L. Baert, Anal. Chem. 64 (1992) 1056.
[154] S. Heron and A. Tchapla, J. Chromatogr. A 848 (1999) 95.
[155] B. Trathnigg and M. Kollroser, J. Chromatogr. A 768 (1997) 223.
[156] W. Miszkiewicz and J. Szymanowski, J. Liq. Chromatogr. & Rel. Technol. 19 (1996) 1013.
[157] J. L. Sims, Chromatographia 53 (2001) 401.
[158] R. Trones, T. Andersen, J. Hunnes and T. Greibrokk, J. Chromatogr. A 814 (1998) 55.
[159] R. Trones, T. Andersen and T. Greibrokk, J. High Resolut. Chromatogr. 22 (1999) 283.
[160] J. N. Alexander, J. Microcol. Sep. 10 (1998) 491.
[161] M. B. O. Andersson and L. G. Blomberg, J. Microcol. Sep. 10 (1998) 249.
[162] Z. Cobb, P. N. Shaw, L. L. Lloyd, N. Wrench and D. A. Barrett, J. Microcol. Sep. 13 (2001) 169.
[163] J. A. Koropchak, S. Sadain, X. Yang, L.-E. Magnusson, M. Heybroek, M. Anisimov and S. L. Kaufman, Anal. Chem. 71 (1999) 386A.
[164] F. S. Deschamps, A. Baillet and P. Chaminade, Analyst 127 (2002) 35.
[165] A. M. Krstulovic, H. Colin and G. A. Guiochon, Adv. Chromatogr. 24 (1984) 83.
[166] D. M. Radzik and S. M. Lunte, CRC Crit. Revs. Anal. Chem. 20 (1989) 317.
[167] G. Horvai and E. Pungor, CRC Crit. Revs. Anal. Chem. 21 (1989) 1.
[168] P. R. Haddad and P. E. Jackson, *Ion Chromatography. Principles and Applications*, Elsevier, Amsterdam, 1990.
[169] P. T. Kissinger and W. E. Heineman, *Laboratory Techniques in Electroanalytical Chemistry*, Dekker, New York, NY, 1995.
[170] J.-G. Chen, S. J. Woltman and S. G. Weber, Adv. Chromatogr. 36 (1996) 273.
[171] W. R. LaCourse, *Pulsed Electrochemical Detection in High-Performance Liquid Chromatography*, Wiley, New York, NY, 1997.
[172] W. W. Buchberger, J. Chromatogr. A 884 (2000) 3.
[173] J. Weiss, *Ion Chromatography*, VCH, Weinheim, 1995.
[174] W. W. Buchberger, Trends Anal. Chem. 20 (2001) 296.
[175] P. Kuban, B. Karlberg, P. Kuban and V. Kuban, J. Chromatogr. A 964 (2002) 227.
[176] H. Yu, Chromatographia 50 (1999) 223.
[177] P. R. Haddad, M. J. Shaw and G. W. Dicinoski, J. Chromatogr. A 956 (2002) 59.
[178] S. Rabin, J. Stillian, V. Barreto, K. Friedman and M. Toofan, J. Chromatogr. 640 (1993) 97.
[179] W. W. Buchberger and P. R. Haddad, J. Chromatogr. A 789 (1997) 67.
[180] T. Brinkmann, C. H. Specht and F. H. Frimmel, J. Chromatogr. A 957 (2002) 99.
[181] P. K. Dasgupta, J. Chromatogr. Sci. 27 (1989) 422.
[182] J. Schultheiss, D. Jensen and R. Galensa, J. Chromatogr. A 880 (2000) 233.
[183] A. Caliamanis, M. J. McCormick and P. D. Carpenter, Anal. Chem. 71 (1999) 741.
[184] Y. Huang, S. Mou and K. Liu, J. Chromatogr. A 832 (1999) 141.
[185] L. M. Nair and R. Saari-Nordhaus, J. Chromatogr. A 804 (1998) 233.
[186] H. Small and J. Riviello, Anal. Chem. 70 (1998) 2205.
[187] R. Saari-Nordhaus and J. M. Anderson, J. Chromatogr. A 956 (2002) 15.
[188] M. Novic, Y. Liu, N. Avdalovic and B. Pihlar, J. Chromatogr. A 957 (2002) 65.
[189] T. Nagatsu and K. Kojima, Trends Anal. Chem. 7 (1988) 21.
[190] J. M. Slater and E. J. Watt, Analyst 119 (1994) 273.
[191] A. Siddiqui and D. C. Shelly, J. Chromatogr. A 691 (1995) 55.

[192] S. R. Wallenborg, K. E. Markides and L. Nyholm, Anal. Chim. Acta 344 (1997) 77.
[193] A. G. Ewing, J. M. mesaros and P. F. Gavin, Anal. Chem. 66 (1994) 527A.
[194] G. Y. Chen and A. G. Ewing, Crit. Rev. Neurobiol. 11 (1997) 59.
[195] O. Niwa, H. Tabei, B. P. Solomon, F. M. Xie and P. T. Kissinger, J. Chromatogr. B 670 (1995) 21.
[196] C. Terashima, H. Tanaka and M. Furuno, J. Chromatogr. A 828 (1998) 113.
[197] Y. C. Lee, J. Chromatogr. A 720 (1996) 137.
[198] D. C. Johnson and W. R. LaCourse, Anal. Chem. 62 (1990) 589A.
[199] D. C. Johnson, D. Dobberpuhl, R. Roberts and P. Vandeberg, J. Chromatogr. 640 (1993) 79.
[200] W. R. LaCourse and C. O. Dasenbrock, Adv. Chromatogr. 38 (1998) 189.
[201] Ch. E. Kientz and U. A. Th. Brinkman, Trends Anal. Chem. 12 (1993) 363.
[202] R. P. W. Scott, C. Little and M. de la Pena, J. Chromatogr. Sci. 38 (2000) 483.
[203] R. P. W. Scott, C. J. Little and M. de la Pena, Chromatographia 53 (2001) S-218.
[204] B. N. Zegers, J. F. C. Debrouwer, A. Poppema, H. Lingeman and U. A. Th. Brinkman, Anal. Chim. Acta 304 (1995) 47.
[205] E. W. J. Hooijschuur, C. E. Kientz and U. A. Th. Brinkman, J. High Resolut. Chromatogr. 23 (2000) 309.
[206] H. L. Peters, K. E. Levine and B. T. Jones, Anal. Chem. 73 (2001) 453.
[207] M. Grotti, P. Rivaro and R. Frache, J. Anal. Atom. Spectrom. 16 (2001) 270.
[208] R. Lobinski, Appl. Spectros. 51 (1997) A260.
[209] R. S. Lobinski, I. R. Pereiro, H. Chassaigne, A. Wasik and J. Szpunar, J. Anal. Atom. Spectrom. 13 (1998) 859.
[210] J. Szpunar, S. McSheehy, K. Polec, V. Vacchina, S. Mounicou, I. Rodriguez and R. Lobinski, Spectrochim. Acta, Part B 55 (2000) 779.
[211] K. Sutton, R. M. C. Sutton and J. A. Caruso, J. Chromatogr. A 789 (1997) 85.
[212] H. Klinkenberg, S. van der Wal, C. de Koster and J. Bart, J. Chromatogr. A 794 (1998) 219.
[213] K. L. Sutton and J. A. Caruso, J. Chromatogr. A 856 (1999) 243.
[214] C. B'Hymer, K. L. Sutton and J. A. Caruso, J. Anal. Atom. Spectrom. 13 (1998) 855.
[215] C. Rivas, L. Ebdon and S. J. Hill, J. Anal. Atom. Spectrom. 11 (1996) 1147.
[216] E. H. Larsen, Spectrochim. Acta, Part B 53 (1998) 253.
[217] J. K. Nicholson, J. C. Lindon, G. B. Scarfe, I. D. Wilson, F. Abou-Shakra, A. B. Sage and J. Castro-Perez, Anal. Chem. 73 (2001) 1491.
[218] L. S. Ahmed, H. Moorehead, C. A. Leitch and E. A. Liechty, J. Chromatogr. B 710 (1998) 27.
[219] J. M. Link and R. E. Synovec, Anal. Chem. 71 (1999) 2700.
[220] D. M. Wieland, M. C. Tobes and T. J. Mangner (Eds.), *Analytical and Chromatographic Techniques in Radiopharmaceutical Chemistry*, Springer-Verlag, New York, 1986.
[221] K. O. Boemsen, J. M. Floeckher and G. J. M. Bruin, Anal. Chem. 72 (2000) 3956.
[222] D. R. Bobbitt and S. W. Linder, Trends Anal. Chem. 20 (2001) 111.
[223] F. Brandt, N. Puster and A. Mannschreck, J. Chromatogr. A 909 (2001) 147.
[224] A. Yamamoto, S. Kodama, A. Matsunaga, K. Hayakawa, Y. Yasui and M. Kitaoka, J. Chromatogr. A 910 (2001) 217.
[225] N. Purdie and K. A. Swallows, Anal. Chem. 61 (1989) 77A.
[226] M. R. Hadley and G. D. Jonas, Enantiomer 5 (2000) 357.
[227] M. Driffield, E. T. Bergstrom, D. M. Goodall, A. S. Klute and D. K. Smith, J. Chromatogr. A 939 (2001) 41.
[228] I. S. Krull (Ed.), *Reaction Detection in Liquid Chromatography*, Dekker, New York, NY, 1986.
[229] J. T. Stewart and W. J. Bachman, Trends Anal. Chem. 7 (1988) 106.
[230] S. M. Lunte, Trends Anal. Chem. 10 (1991) 97.
[231] P. Jones, Analyst 125 (2000) 803.
[232] Y. Ohkura, M. Kai and H. Nohta, J. Chromatogr. B 659 (1994) 85.
[233] D. Kutlan, P. Presits and I. Molnar-Perl, J. Chromatogr. A 949 (2002) 235.
[234] S. Honda, J. Chromatogr. A 720 (1996) 183.

[235] Z. El Rassi (Ed.), *Carbohydrate Analysis. High Performance Liquid Chromatography and Capillary Electrophoresis*. Elsevier, Amsterdam, 1995.
[236] B. D. McGarvey, J. Chromatogr. B 659 (1994) 243.
[237] W. H. Newsome, B. P. Y. Lau, D. Ducharme and D. Lewis, J. AOAC Int. 78 (1995) 1312.
[238] S. R. Ruberu, W. M. Draper and S. K. Perera, J. Agric. Food Chem. 48 (2000) 4109.
[239] H. Engelhardt and U. D. Neue, Chromatographia 15 (1982) 403.
[240] A. H. M. Scholten, U. A. Th. Brinkman and R. W. Frei, Anal. Chem. 54 (1982) 1932.
[241] R. S. Deelder, A. T. J. M. Kuijpers and J. H. M. Van den Berg, J. Chromatogr. 255 (1983) 545.
[242] L. Dalgaard, Trends Anal. Chem. 5 (1986) 185.
[243] G. M. Ware, G. W. Chase, R. R. Eitenmiller and A. R. Long, J. AOAC Int. 83 (2000) 957.
[244] S. Waiz, B. M. Cedillo, S. Jambunathan, S. G. Hohnbolt, P. K. Dasgupta and D. K. Wolcott, Anal. Chim. Acta 428 (2001) 163.
[245] J. R. Poulsen, K. S. Birks, M. S. Grandelman and J. W. Birks, Chromatographia 22 (1986) 231.
[246] M. Kramer and H. Engelhardt, J. High Resolut. Chromatogr. 15 (1992) 24.
[247] O. Kuhlmann and G. J. Krauss, J. Pharm. Biomed. Anal. 16 (1997) 553.
[248] M. Lores, O. Cabaleiro and R. Cela, Trends Anal. Chem. 18 (1999) 392.
[249] P. F. Garcia-Borregon, M. Lores and R. Cela, J. Chromatogr. A 870 (2000) 39.
[250] M. Lores, C. M. Garcia and R. Cela, J. Chromatogr. A 724 (1996) 55.
[251] H. Engelhardt, J. Meister and P. Kolla, Chromatographia 35 (1993) 5.
[252] T. C. Schmidt, D. Meinzer, L. Kaminski, E. von Low and G. Stork, Chromatographia 46 (1997) 501.
[253] R. Gatti, M. G. Gioia, A. M. Di Pietra and M. Cini, J. Chromatogr. A 905 (2001) 345.
[254] E. S. Yeung and W. G. Kuhr, Anal. Chem. 63 (1991) 275A.
[255] F. Steiner, W. Beck and H. Engelhardt, J. Chromatogr. A 738 (1996) 11.
[256] R. P. W. Scott, C. G. Scott and P. Kucera, Anal. Chem. 44 (1972) 100.
[257] T. Takeuchi and T. Miwa, J. High Resolut. Chromatogr. 22 (1999) 609.
[258] T. Takeuchi, K. Kawai, Y. Kitamaki and T. Miwa, Chromatographia 52 (2000) 63.
[259] J. V. Goodpaster and V. L. McGuffin, Anal. Chem. 73 (2001) 2004.

# Chapter 6

# Thin-Layer Chromatography

## 6.1 INTRODUCTION

Thin-layer chromatography (TLC) can trace its origins to the introduction of drop chromatography in the late 1930s [1-4]. A microscope slide was covered with a layer of aluminum oxide, on which one drop of extract was spotted, followed by dropwise addition of solvent onto this spot. Separated substances were visualized as circular zones radiating from the original spot center. For analytical applications drop chromatography offered a faster, more convenient and more powerful separation tool than conventional column chromatography as originally described by Tswett. Thin-layer chromatography as we know it today was not established until the 1950s, in large part due to the efforts of Kirchner and Stahl [5,6]. These workers devised standardized procedures to improve the separation performance and reproducibility of thin-layer chromatography, paving the way for its commercialization, as well as contributing many new applications. The 1970s saw the introduction of fine particle layers and associated instrumentation for their correct use [7-9]. In this form thin-layer chromatography became known as high performance TLC, instrumental TLC, or modern TLC to distinguish it from its parent, now generally referred to as conventional TLC. Modern TLC did not displace conventional TLC from laboratory studies and the two approaches coexist today because of their complementary features. Conventional thin-layer chromatography can provide a quick, inexpensive, flexible and portable method for monitoring synthetic reactions and similar applications. Virtually no instrumentation is required and the method can be learned quickly. On the other hand, modern thin-layer chromatography is characterized by the use of more expensive fine particle layers for faster and more efficient separations, and requires the use of instruments for convenient (automated) sample application, development and detection [1,9-11]. Conventional thin-layer chromatography is considered a qualitative or semi-quantitative method while modern thin-layer chromatography provides accurate and precise quantitative results based on *in situ* measurements and a record of the separation in the form of a chromatogram. Since the 1970s modern thin-layer chromatography has evolved by technical improvements to layers, instrumentation and increased automation facilitated by computer-controlled devices. Major innovations include automated multiple development, interfaces to spectroscopic instruments (sections 9.2.2.5 and 9.3.1.4) and emerging techniques for video densitometry.

Thin-layer chromatography is a type of liquid chromatography in which the stationary phase is in the form of a thin layer on a flat surface rather than packed into a tube (column). It is a member of a family of techniques that include some types of electrophoresis and paper chromatography more generally referred to as planar chromatography. Since we will not discuss electrophoresis here, and since thin-layer chromatography has virtually superseded paper chromatography, we will confine ourselves to a discussion of thin-layer chromatography.

In the basic experiment, the sample is applied to the layer as a spot or band near to the bottom edge of the layer. The separation is carried out in a closed chamber by either of two mechanisms. By contacting the bottom edge of the layer with the mobile phase, which advances through the layer by capillary forces, or by forcing the mobile phase through the layer at a controlled velocity by applying external pressure or other means. A separation of the sample results from the different rates of migration of the sample components in the direction traveled by the mobile phase. After development and evaporation of the mobile phase, the sample components are separated in space; their position and quantity being determined by visual evaluation or densitometry.

## 6.2 ATTRIBUTES OF LAYERS AND COLUMNS

Separations by column liquid chromatography (HPLC) and thin-layer chromatography occur essentially by the same physical processes. The two methods are often considered competitors when it would be more realistic to consider them complementary techniques. The favorable attributes of thin-layer chromatography providing for its co-existence alongside column liquid chromatography are summarized in Table 6.1 [1].

The elution mode is commonly used for column liquid chromatography and the development mode for thin-layer chromatography. Each sample component must travel the complete length of the column in column liquid chromatography and the total separation time is determined by the time required for the slowest moving component to reach the detector. While for thin-layer chromatography the total time for the separation is the time required for the solvent front to migrate a predetermined distance, and is independent of the migration distance of the sample components. Excessively retained components extend the separation time in column liquid chromatography while components accumulated at the head of the column are completely eluted. If this is not possible then permanent alteration of the properties of the column may occur, eventually making the column useless for the separation. Thin-layer chromatography plates are disposed of at the conclusion of each separation and are immune from this problem. Since the total sample occupies the chromatogram in thin-layer chromatography the integrity of the analysis is guaranteed but can only be implied in column chromatography.

Table 6.1
Attributes of thin-layer chromatography

| Attribute | Application |
|---|---|
| Separation of samples in parallel | •Low-cost analysis and high-throughput screening of samples requiring minimal sample preparation. |
| Disposable stationary phase | •Analysis of crude samples (minimizing sample preparation requirements)<br>•Analysis of a single or small number of samples when their composition and/or matrix properties are unknown<br>•Analysis of samples containing components that remain sorbed to the separation medium or contain suspended microparticles |
| Static detection | •Samples requiring postchromatographic treatment for detection<br>•Samples requiring sequential detection techniques (free of time constraints) for identification or confirmation |
| Storage device | •Separations can be archived<br>•Separations can be evaluated in different locations or at different times<br>•Convenient fraction collection for multimodal column/layer chromatography |
| Sample integrity | •Total sample occupies the chromatogram not just that portion of the sample that elutes from the column. |

In addition, the use of the elution mode in column liquid chromatography means that separations must be performed sequentially. The time required to analyze a group of samples is equal to the time to separate one sample, including column re-equilibration time, multiplied by the number of samples in the group. Thin-layer chromatography allows samples to be separated in parallel with the potential for a significant reduction in the total separation time. Depending on the plate size, between 18 and 72 samples and standards can be separated simultaneously in the same time required to separate a single sample. Some of this advantage is lost because it is easier to automate column liquid chromatography, and unattended overnight operation is possible. Whereas for thin-layer chromatography the individual steps of sample application, development and detection can be automated, but manual intervention is required to move the plate from station to station. For thin-layer chromatography, it is also necessary to include the time for sample application and scanning of the plate. Even after making suitable allowances for these points, thin-layer chromatography generally provides a significant reduction in separation time.

At the end of the separation in thin-layer chromatography, the mobile phase is evaporated and the separation becomes immobilized. Detection, therefore, takes place in the presence of the stationary phase, is independent of time constraints, and all separated zones are simultaneously accessible. This provides flexibility in the choice of detection strategies and even the possibility of archiving separations for evaluation by different detection processes applied at different times or locations. Several detection techniques can be applied sequentially, as long as they are nondestructive. For optical detection, the minimum detectable quantities are similar for both column liquid and thin-layer chromatography with, perhaps, a slight advantage for column liquid chromatography. Direct comparisons are difficult because of the differences in detection variables and how these are optimized. Detection in thin-layer chromatography, however, is generally limited to optical detection without the equivalent of refractive index and electrochemical detection available for column liquid chromatography. The ease of postchromatographic reactions used to enlarge the application range of optical detection in thin-layer chromatography largely offsets this disadvantage.

In the light of the above discussion, thin-layer chromatography is most effective for the low-cost analysis of simple mixtures, especially when a large number of samples require analysis. For the rapid analysis of samples requiring minimum sample cleanup or where thin-layer chromatography allows a reduction in the number of sample preparation steps (e.g. the analysis of samples containing components that remain sorbed to the stationary phase or contain suspended microparticles). For the analysis of substances with poor detection characteristics requiring postchromatographic treatment for detection. In addition, it is the most suitable technique for determining general sample composition, since all sample components are located in the chromatogram.

Traditionally, thin-layer chromatography has been used in large-scale surveillance programs. These include the identification of drugs of abuse and toxic substances in biological fluids in forensic toxicology [12-15]. The identification of unacceptable levels of residues from drugs used to prevent disease or promote growth in farm animals [16-19]. To ensure a safe water supply by monitoring crop-protecting agents used in modern agriculture [20,21]. To ensure conformity with the label declaration of pharmaceutical products [22]. In these applications, thin-layer chromatography is often used with other methods in a pyramid strategy. This strategy employs a screening step (thin-layer chromatography) to identify suspect samples and a confirmation step (the most suitable analytical method) to establish the contaminant level in suspect samples. The benefits of the pyramid approach are lower costs and an increase in the number of samples processed to identify violated samples [1]. Thin-layer chromatography is selected for the screening step because: (1) single use of the stationary phase minimizes sample preparation requirements; (2) parallel separations enhance sample throughput; (3) ease of postchromatographic derivatization improves selectivity and specificity for the analysis; and (4) several screening protocols for different analytes can be carried out simultaneously.

Thin-layer chromatography remains one of the main methods for class fractionation and speciation of lipids [23,24] and is used increasingly to determine the botanical origin, potency, and flavor potential of herbs and spices [25-27]. In the pharmaceutical industry, it is used for the analysis of complex and dirty samples with poor detection characteristics and for stability and content uniformity testing [28-31]. It continues to be widely used in the standardization of plant materials used as traditional and modern medicines. In addition, it retains an historic link with the characterization of dyes and inks and the control of impurities in industrial chemicals.

In other cases, column liquid chromatograph is the preferred method of separation, particularly if a large plate number is required for the separation or the sample preparation time is long compared with the separation time. Separations by size-exclusion and ion-exchange chromatography are usually easier by column liquid chromatography. Since there are few suitable layers for the separation of biopolymers by thin-layer chromatography, column liquid chromatography and electrophoresis are commonly used for these applications. Finally, column liquid chromatography is favored for trace analysis using selective detectors unavailable to thin-layer chromatography.

## 6.3 THEORETICAL CONSIDERATIONS

A unique feature of thin-layer chromatography is the presence of a third phase, the vapor phase, in contact with the mobile and stationary phases. Alterations in the composition and velocity of the mobile phase moving through the layer induced by the vapor phase are possible in ways that are not observed for column liquid chromatography. Also, the mobile phase normally penetrates a dry layer in thin-layer chromatography by capillary action. The flow of solvent at the developing front is generally unsaturated and the speed with which the front moves is dependent on experimental and environmental conditions [32-35]. Capillary forces are stronger in the narrow interparticle channels, leading to a more rapid advance of the mobile phase. Larger pores below the solvent front are filled at a slower rate resulting in an increased thickness of the mobile phase layer. If the vapor phase and mobile phase are not in equilibrium, evaporation will cause a loss of mobile phase from the layer surface and a decrease in the solvent front velocity. On the other hand, the dry layer ahead of the solvent front progressively adsorbs vapor, filling some of the pores and interparticle channels, and increasing the apparent velocity with which the solvent front migrates. During the chromatographic process a solvent composition gradient is produced as the mobile phase moves through the sorbent due to selective adsorption by the stationary phase of the solvent component with the higher affinity for the stationary phase. Gaining satisfactory control over the above processes in large volume chambers is almost impossible. Various kinds of sandwich chambers, which either eliminate or minimize contact of the plate surface with the vapor phase, offer reasonable control of the mobile phase velocity.

### 6.3.1 Retardation Factor

The fundamental parameter used to characterize the position of a sample zone in a thin-layer chromatogram is the retardation factor, or $R_F$ value. It represents the ratio of the distance migrated by the sample compared to the distance traveled by the solvent front, and for linear development is given by Eq. (6.1)

$$R_F = Z_X / (Z_f - Z_o) \tag{6.1}$$

where $Z_X$ is the distance traveled by the sample from its origin, $(Z_f - Z_o)$ the distance traveled by the mobile phase from the sample origin, $Z_f$ the distance traveled by the solvent front measured from the mobile phase entry position (the solvent level at the start of the separation), and $Z_0$ the distance from the sample origin to the position used as the origin for the mobile phase. The boundary conditions for $R_F$ values are $1 \geq R_F \geq 0$. When $R_F = 0$, the spot does not migrate from the origin, and for $R_F = 1$, the spot is unretained by the stationary phase and migrates with the solvent front. Although $R_F$ values are widely quoted, they are difficult to determine accurately [32,36]. Systematic errors result from the difficulty in locating the exact position of the solvent front. If the adsorbent layer, mobile phase and vapor phase are not in equilibrium then condensation of the vapor phase or evaporation of the mobile phase in the region of the solvent front will give an erroneous $R_F$ value. Evaporation of mobile phase from the surface of the layer in unsaturated chambers generally results in higher $R_F$ values. The reproducibility of $R_F$ values in unsaturated chambers may be poor and concave solvent fronts formed resulting in higher $R_F$ values for samples nearest to the edges of the layer. The thermodynamic solvent front, which may be slightly lower than the visible solvent under conditions of unsaturated flow, can be determined by using an unretained substance as a solvent-front marker [37]. Not only does the level of saturation influence the reproducibility of $R_F$ values but also can result in changes in the migration order [36]. To improve the reproducibility of $R_F$ values determined at different times or locations, the saturation grade of the developing chamber is used to interconvert observed $R_F$ values between different chamber designs [36]. When the analytes are available as standards, it is common practice to separate the standards and samples in the same system for identification purposes. In surveillance programs, the simultaneous separation of appropriate standard substances is used to improve the certainty of identification by correcting observed $R_F$ values to standard $R_F$ values for automated library searches [12-15]. Using the mean list method, for example, all substances that migrate in a $R_F$ window that might be confused, are ranked and compared across a number of separation systems. If the separation systems are complementary, the list of possible substances that might be confused will become shorter as an increasing number of substances fall outside the identification window for the unknown. Eventually only a handful of possible substances remain on the list. At which point suitable selective separation and spectroscopic techniques are used to confirm the identification of the unknown.

$R_F$ values are generally calculated to two decimal places. Some authors prefer to tabulate values as whole numbers, as $hR_F$ values equivalent to 100 $R_F$. The $R_F$ value is not linearly related to the distribution properties of the separation system. The $R_M$ value is used in studies that attempt to correlate migration properties to solute structure. The $R_M$ value is equivalent to the ratio of the residence time of the solute in the stationary and mobile phases, and is formally equivalent to the retention factor (log k) in column liquid chromatography. It is calculated from the $R_F$ value by $R_M$ (or log k) = $\log [(1 - R_F) / R_F]$.

### 6.3.2 Flow Through Porous Layers

The common methods of mobile phase transport through the layer are capillary action, forced flow, and electroosmosis. Ease of implementation results in capillary flow being the dominant transport mechanism, although it is also the most restrictive. Forced flow has a number of advantages over capillary flow but requires sophisticated instruments and is less widely used. Electroosmotic flow has shown promise as a transport mechanism, but there is confusion over conditions required for its utilization. For now, it remains a curiosity in need of further development.

*6.3.2.1 Capillary Flow*

The force driving mobile phase migration in thin-layer chromatography is the decrease in the free energy of the solvent as it enters the porous structure of the layer; the transport mechanism is a result of capillary action [32,33]. It is an empirical fact that in the absence of a significant exchange of solvent flux with the vapor phase the position of the solvent front with respect to time is adequately represented by the simple quadratic expression $(Z_f)^2 = \kappa t$, and after differentiation, the velocity of the solvent front by $u_f = \kappa / 2Z_f$. Where $Z_f$ is the distance of the solvent front position above the solvent level in the developing chamber, $\kappa$ the velocity constant ($cm^2 / s$), t the time from contacting the layer with the solvent, and $u_f$ the solvent front velocity. In the absence of equilibrium, rather complex correction factors must be applied to the quadratic equation relating $Z_f$ to the development time [32,33]. The undesirable feature of capillary flow, namely, decreasing mobile phase velocity with increasing migration distance, is apparent from the form of the solvent-front velocity expression. The outcome is longer separation times and a reduced separation potential.

The velocity constant, $\kappa$, is related to the experimental conditions by equation (6.2)

$$\kappa = 2\, K_o\, d_p\, (\gamma / \eta) \cos \theta \tag{6.2}$$

where $K_o$ is the permeability constant of the layer, $d_p$ the average particle diameter, $\gamma$ the surface tension of the mobile phase, $\eta$ the viscosity of the mobile phase, and $\theta$ the contact angle. The permeability constant is a dimensionless constant which takes into account the profile of the external pore size distribution, the effect of porosity on the permeability of the layer, and the ratio of the bulk liquid velocity to the solvent front velocity. Experimental values for $K_o$ vary for different layers. Typical values for precoated layers, however, fall into the range 0.001 to 0.002 and are similar to values for slurry-packed columns [38].

Equation (6.2) indicates that the velocity constant should increase linearly with the average particle size resulting in a solvent-front velocity that should be larger for coarse-particle layers than for fine-particle layers. Also, from Eq. (6.2) we see that the velocity constant varies linearly with the ratio of the surface tension of the solvent to its viscosity. Thus, solvents which maximize this ratio (and not just optimize one of the parameters) are preferred for thin-layer chromatography. The contact angle for most organic solvents on silica gel and polar chemically bonded layers is generally close to zero ($\cos \theta = 1$). This is not the case for chemically bonded reversed-phase layers containing long-chain, alkyl groups. For these layers, the contact angle of the mobile phase increases rapidly with the water content and, when the water content is about 30-40%, $\cos \theta$

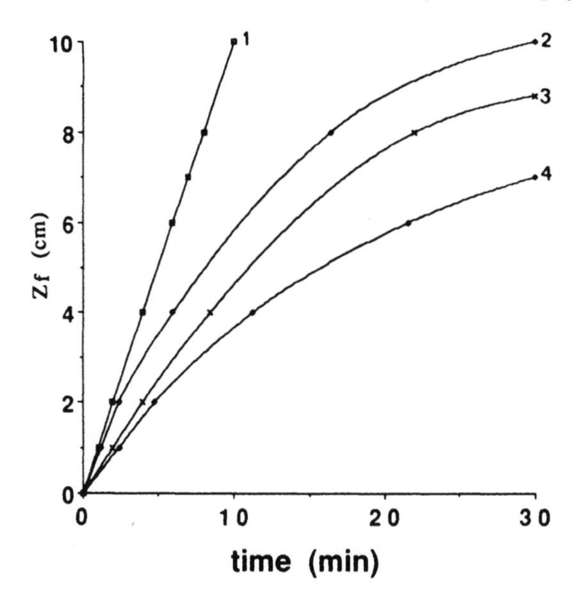


Figure 6.1. Relationship between the solvent-front position and time for a forced flow separation (1) and capillary flow separations with an exposed layer in a saturated chamber (2), a covered layer (sandwich chamber) (3), and an exposed layer in an unsaturated atmosphere (4). (From ref. [33]; ©Research Institute for Medicinal Plants).

becomes less than 0.2-0.3 [39]. Separations dependent on capillary flow are just about impossible under these conditions. For reasonable compatibility with aqueous mobile phases, hydrophobic layers with less silanization and/or larger particle size are generally used for reversed-phase thin-layer chromatography.

The above discussion is applicable to layers unperturbed by the presence of a vapor phase, such as in a sandwich-type-developing chamber. In practice, most separations are performed in large volume chambers in the presence of a vapor phase. It is almost impossible to saturate these chambers and temporal and spatial vapor equilibrium is unlikely to exist. Two opposing phenomena are expected to influence the rate of mobile phase migration. Vaporization of the mobile phase from the wetted layer is reasonably expected to depend on the wetted surface area of the plate and the vapor pressure of the solvents in the chamber. The loss of mobile phase from the layer will result in a reduction of the mobile phase velocity from that indicated by Eq. (6.2), Figure 6.1 [34,40]. When a dry layer is placed in the developing chamber, it progressively adsorbs mobile phase vapors. The pores of the dry layer ahead of the solvent front fill slowly with adsorbed vapor and the apparent porosity of the layer diminishes. Since the porosity of the layer decreases, the velocity constant increases slowly with time. The effect of vaporization is generally small if the chamber atmosphere is close to saturation. In this case, adsorption of mobile phase vapors by the dry layer will tend to dominate. Thus, the mobile phase velocity in a large-volume chamber will tend to be greater than that given by Eq. (6.2) and increases with time, if the layer is not conditioned in the chamber atmosphere prior to the start of development.

*6.3.2.2 Forced Flow*

Forced flow enables the mobile phase velocity to be optimized without regard to the deficiencies of capillary flow systems. Rotational planar chromatography uses centrifugal force generated by spinning the layer about a central axis, to drive the mobile phase through the layer [41,42]. The mobile phase velocity is a function of the rotation speed and the rate at which the mobile phase is supplied to the layer. Since the layer is not enclosed, the ultimate velocity of the solvent front is limited by the amount of solvent that can be kept within the layer without floating over the surface. At high rotation speeds, the velocity of the solvent front is approximately constant in the linear development mode. Rotational planar chromatography is used for preparative-scale applications (section 11.2) more so than analytical separations.

An alternative approach to forced flow is to seal the layer with a flexible membrane or an optically flat, rigid surface under hydraulic pressure, and to deliver the mobile phase to the layer by a pump [9,41,43-46]. Adjusting the solvent volume delivered to the layer optimizes the mobile phase velocity. In the linear development mode, the mobile phase velocity ($u_f$) will be constant and the position of the solvent front ($Z_f$) at any time (t) after the start of development is described by $Z_f = u_f t$. The mobile phase velocity no longer depends on the contact angle and solvent selection is unrestricted for reversed-phase layers in forced flow, unlike capillary flow systems.

When a liquid is forced though a dry layer of porous particles sealed from the external atmosphere, the air displaced from the layer often form a second front (beta front), moving behind the mobile phase-air front (alpha front) [47-49]. The beta front usually has an irregular (wavy) shape. During the liquid-air displacement process, some air is displaced instantaneously and escapes ahead of the front. The remainder is displaced at a slower rate and leaves the layer by dissolution in the mobile phase, or as microbubbles moving with the mobile phase. The solubility of air in the mobile phase depends on the applied pressure, and at some critical pressure, all of the gas will dissolve and the layer becomes completely wetted by the mobile phase. The space between the alpha and beta fronts is referred to as the disturbing zone. It can be distinguished from the completely wetted region of the layer by its apparent optical density. Solutes moving in the disturbing zone, or passed over by it, are often distorted and difficult to quantify by scanning densitometry. The disturbing zone is minimized or eliminated by predevelopment of the layer with a weak mobile phase in which the sample does not migrate. This dislodges the trapped air from the layer before starting the separation. Increasing the local pressure by using a backpressure regulator increases the solubility of air in the mobile phase and provides an alternative approach to minimize the disturbing zone.

Multiple solvent fronts are also observed for both forced flow and capillary flow systems with mobile phases containing solvents of different elution strength [32,49]. As the mobile phase moves through the layer, it becomes depleted in component with greatest affinity for the stationary phase. Eventually a secondary front is formed that separates the equilibrium mobile phase composition from the mobile phase composition now totally depleted in the solvent selectively adsorbed by the stationary phase. For a mobile phase consisting of n solvents, as many as n+1 solvent fronts are possible. This process is referred to as solvent demixing and adds to the complexity of method development. Solvent demixing results in sample components moving in regions of different solvent strength and selectivity with retention that is no longer easily related to the bulk mobile phase composition. Sample zones close to a secondary front are often focused into narrow bands with a different shape to neighboring zones, which allows demixing conditions to be identified. Solvent demixing is intensified in developing chambers with an unsaturated atmosphere.

#### *6.3.2.3 Electroosmotic Flow*

Electroosmotic flow has emerged as a viable alternative transport mechanism to pressure-driven flow in column chromatography (section 8.4.1). Benefits include a plug-flow profile (reduced transaxial contributions to zone broadening) and a mobile phase velocity that is independent of the column length and particle size (assuming Joule heating is not a limiting factor). The simple experimental arrangement for generating electroosmotic flow and declining interest in forced flow are responsible for the current interest in electroosmotic-driven flow as a means of overcoming the limitations of capillary flow in thin-layer chromatography [11,50].

The current status of electroosmotic-driven flow in thin-layer chromatography (planar electrochromatography) is probably more confusing than reassuring, although recent studies have brought some enlightenment to this technique [50-54]. Early studies of electroosmotic flow in vertically mounted layers are now believed, for the most part, to be the result of thermal effects [51]. Enhanced flow resulting from forced evaporation of the mobile phase from a solvent-deficient region at the top of the layer. In horizontal developing chambers, as well as electroosmotic flow in the direction of migration there is also flow to the surface of the layer. Joule heating offsets this effect by evaporation of mobile phase from the layer surface. Under conditions of excessive wetting or drying of the layer, degradation of the separation quality occurs [53,54]. For acceptable separation quality in reversed-phase separations, voltage and buffer concentration should be optimized to minimize either excessive wetting or drying of the layer.

True electroosmotic flow has been demonstrated in horizontally mounted layers at modest field strengths (< 1 kV / cm) with mobile phases of high dielectric constant. Still unclear is which solvents can be used, the need for prewetted layers and ions as current carriers, the effect of local heating on zone profiles, and the effect of binder chemistry on flow, mass transfer, and thermal effects. Compared with capillary flow faster separations have been demonstrated, but the influence of flow velocity on efficiency was only treated in a qualitative sense. So far no comprehensive analysis of the kinetic properties of separations under conditions of electroosmotic flow have been performed in thin-layer chromatography, and too little information is available to warrant formulation of a theoretical model until mechanistic details are clarified.

### 6.3.3 Zone Broadening and the Plate Height Equation

A discussion of zone broadening in thin-layer chromatography is inherently more complex than for column liquid chromatography. The reasons for this are due to both the practice of the technique and the limitations of the subsequent theoretical treatment [11,32,55-59]. Visual detection cannot be relied upon to determine zone dimensions since the eye is incapable of accurately identifying zone boundaries at low concentration and instrumental techniques are essential. The size of developed zones in thin-layer chromatography are normally in the 2-10 mm range and sample application zones about 1-2 mm. Consequently, the size of the starting zone is never small compared to the size of developed zones. The contribution of the starting zone length to those of the separated zones can be removed by consideration of their variance. The determination of zone variance is straightforward for developed zones but presents some difficulty for the starting zone [60]. On first contacting the sample zone the advancing solvent front can cause a reshaping of the zone and, since its leading edge is probably not fully saturated, all the pores holding sample will not be filled simultaneously. Consequently, adsorbed sample may not be displaced instantaneously from the sorbent surface, and localized solvent saturation may limit the rate of solute dissolution in the mobile phase. These factors will generally prevent an accurate value for the starting zone dimensions relevant to the separation from being obtained.

Zones in modern thin-layer chromatography are generally symmetrical and their profiles recorded by scanning densitometry are Gaussian to a good approximation. The average plate height ($H_{obs}$) and plate number ($N_{obs}$) can be calculated directly from the chromatogram as

$$H_{obs} = w^2 / a Z_X \quad (6.3)$$

$$N_{obs} = a (Z_X / w)^2 \quad \text{and} \quad Z_X = R_F (Z_f - Z_o) \quad (6.4)$$

where w is a parameter describing the peak width and $a$ is an appropriate scaling factor for a Gaussian model (section 1.5). When w is the peak width at the base $a$ has the value 16, and when w is the peak width at half height, $a$ is 5.54. In thin-layer chromatography all separated zones migrate different distances, which is not the case for column chromatography. Consequently, separated zones only experience those theoretical plates through which they travel and, therefore, the average plate height and plate number depend on the zone migration distance. Also, an average value is used to indicate that the mobile phase velocity for capillary flow is variable throughout the chromatogram and individual zones migrate through regions of different local efficiency. Thus, the average plate height and the observed plate height in the chromatogram are identical. For the same reason, the average plate height depends

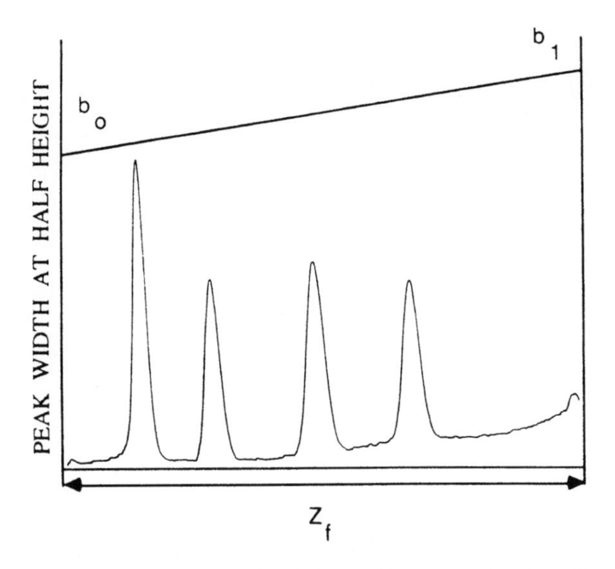


Figure 6.2. Change in peak width at half height as a function of migration distance for a test mixture on a high performance layer. Linear regression analysis is used to obtain values for $b_0$ and $b_1$.

on the distance between the sample application position and the mobile phase entry position or solvent level for the layer ($Z_o$). As $Z_o$ increases, sample zones are forced to migrate with a lower mobile phase velocity resulting in additional zone broadening [59].

Kaiser proposed an alternative method to determine the plate height based on linear extrapolation of the peak width at half height (given the symbol b) as a function of ($Z_f - Z_o$) [7,57]. The hypothetical expected peak widths at half height corresponding to $R_F$ = 0 ($b_0$) and $R_F$ = 1 ($b_1$) are obtained as indicated in Figure 6.2. The real plate number ($N_{real}$) and the real plate height ($H_{real}$) can then be defined according to equations (6.5) and (6.6). This approach recognizes the importance of the starting zone dimension on the plate height value and provides a simple means to calculate plate height values for any $R_F$ value independent of the location of substances in the chromatogram. For the purposes of standardization $N_{real}$ and $H_{real}$ are usually quoted for $R_F$ = 0.5 or 1.0.

$$N_{real} = 5.54\ [Z_X / (b_X - b_0)]^2 \quad (6.5)$$

$$H_{real} = (b_X - b_0)^2 / 5.54\ Z_X \quad (6.6)$$

Ideally the rest diffusion value, $b_0$, should be small compared to $b_1$ to maximize $N_{real}$. In practice the extrapolated value for $b_0$ is somewhat independent of the size of the sample application zone for normal zone sizes and depends significantly on the quality of the layer. The rest diffusion value is probably not an accurate estimate of the starting zone size at the time of initial migration [59].

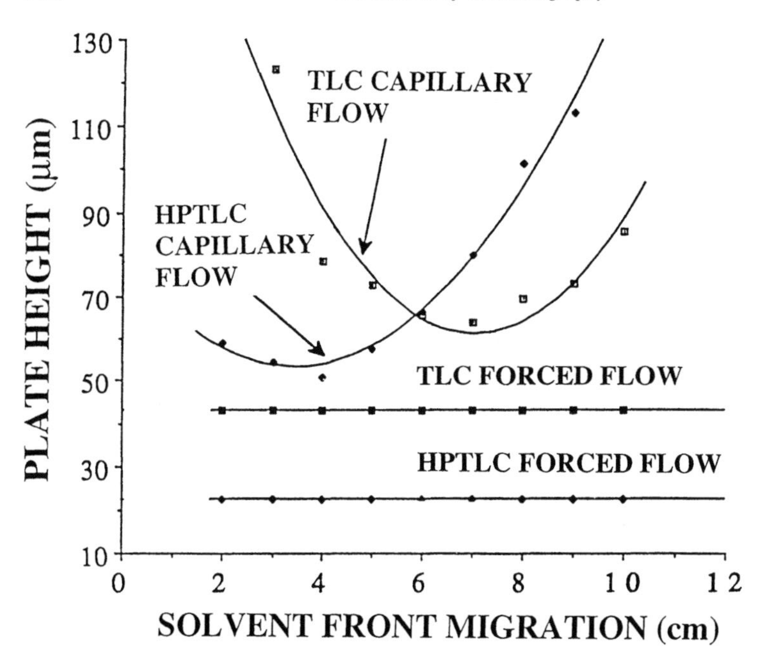


Figure 6.3. Variation of the average plate height as a function of the solvent-front migration distance for conventional and high performance layers using capillary flow and forced flow at the optimum mobile phase velocity. (From ref. [61]; ©Elsevier).

Capillary flow results in a mobile phase velocity that is spatially dependent (declines with the solvent-front migration distance) with a maximum velocity at any position on the layer that is less than the velocity required for optimum performance [11,59,61]. For common solvents the mobile phase velocity is typically in the range 0.02 to 0.005 cm/s while the optimum mobile phase velocity for high performance layers is about 0.05 cm/s [62]. A plot of the average plate height as a function of the solvent-front migration distance, Figure 6.3, reveals several important features of zone broadening with capillary flow. There is a dominant relationship between the solvent-front migration distance, the average particle size for the layer, and the separation performance. High performance layers of small particle size afford compact zones if the solvent-front migration distance does not exceed about 5-6 cm. When solvent-front migration distances are longer than this, layers of larger average particle size (conventional layers) are more efficient. This contrary finding is easily explained by the relative permeability of the layers. The mobile phase velocity for the high performance layer declines rapidly with the solvent-front migration distance until eventually zone broadening exceeds the rate of migration of the zone center. It is futile to use longer solvent-front migration distances than 5-6 cm for high performance layers for this reason. For the more permeable conventional layers the mobile phase velocity is higher for longer distances. As expected, the minimum in the plate height is shifted to longer

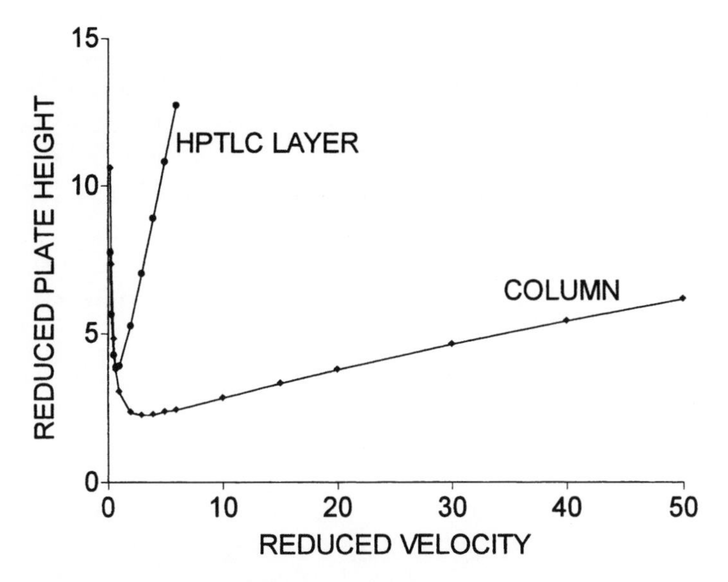


Figure 6.4. Plot of the reduced plate height against the reduced mobile phase velocity for a high performance layer and column. (From ref. [1]; ©Elsevier)

solvent-front migration distances for the conventional layer, but the intrinsic efficiency, measured by the minimum plate height, is more favorable for the high performance layer. When the development distance is optimized the separation performance of conventional and high performance layers are not very different. The virtues of the high performance layers are that it requires a shorter migration distance to achieve a given efficiency resulting in faster separations and more compact zones, which are easier to detect by scanning densitometry.

For forced flow separations a constant plate height independent of the solvent-front migration distance is obtained, Figure 6.3. The minimum plate height for capillary flow is always greater than the minimum for forced flow. This is an indication that the limited range of capillary flow velocities is inadequate to realize the optimum kinetic performance for the layers. At the mobile phase optimum velocity, forced flow affords more compact zones and shorter separation times compared with capillary flow. As expected the intrinsic efficiency increases with a reduction of the average particle size for the layer.

The variation of the reduced plate height (section 1.5.3) as a function of the reduced mobile phase velocity for a packed column and high performance layer is illustrated in Figure 6.4 [1,62]. The optimum reduced mobile phase velocity is shifted to a lower value compared with the column and the minimum in the reduced plate height ($\approx$ 3.5) is higher than typical values for a good column ($\approx$ 2.0-2.5). Also, at higher reduced mobile phase velocities the reduced plate height for the layer is significantly larger than for the column. Resistance to mass transfer is significantly higher for the layer than for a good

Table 6.2
Performance characteristics of forced flow thin-layer chromatography
Assumptions: viscosity = $3.5 \times 10^{-4}$ N.s/m$^2$ and solute diffusion coefficient = $2.5 \times 10^{-9}$ m$^2$/s. (h = reduced plate height, ν = reduced mobile phase velocity, ϕ = flow resistance parameter, and $d_p$ = average particle size)

| Development time (min) | Pressure drop (atm) | $N_{max}$ | Development length (cm) |
|---|---|---|---|
| *HPTLC (optimum conditions* h = 3.75, ν = 0.8, ϕ = 800 and $d_p$ =6 μm) | | | |
| 4 | 2.1 | 3,550 | 8 |
| 9 | 4.7 | 8,000 | 18 |
| 25 | 12.9 | 22,200 | 50 |
| 50 | 25.8 | 44,400 | 100 |
| *HPTLC (fast development option h = 9 and ν = 5)* | | | |
| 0.6 | 12.9 | 1,480 | 8 |
| 1.4 | 29.1 | 3,330 | 18 |
| 4.0 | 80.7 | 9,250 | 50 |
| 8.0 | 161.0 | 18,500 | 100 |
| *HPTLC ($d_p$ = 3 μm; other parameters as in (a))* | | | |
| 2.0 | 16.5 | 7,610 | 8 |
| 4.5 | 37.2 | 17,100 | 18 |
| 12.5 | 103 | 47,600 | 50 |
| 25.0 | 207 | 95,200 | 100 |
| *Conventional TLC (h = 4.5 and ν = 0.8,* ϕ = 600 and $d_p$ = 9 μm) | | | |
| 6 | 0.44 | 1,980 | 8 |
| 13.5 | 1.03 | 4,450 | 18 |
| 37.5 | 2.9 | 12,350 | 50 |
| 75 | 5.7 | 24,700 | 100 |

column. Consequently, separations by forced flow will be slower than those achieved with typical columns, and fast separations at high flow rates will be less efficient.

Some expectations for forced flow separations are summarized in Table 6.2 [1,50]. For a normal development distance of 18 cm a modest increase in performance (a maximum of 8000 theoretical plates) in a credible time of 9 minutes is achieved compared with typical results for capillary flow (< 5000 theoretical plates in about 25-45 min). Really significant increases in efficiency are achieved only by the use of longer layers at the expense of separation time. This can be achieved by the serial coupling of conventional size layers arranged in a stack with a special tail-to-head vertical connection between layers [46,63,64]. The optimum mobile phase velocity is sufficiently low that the pressure drop remains modest compared with typical coupled column systems for liquid chromatography. Fast separations on high performance layers, result in low efficiency, and high pressures are required if long development distances are used. These conditions are not useful for most separations, especially because layers enable separations to be performed in parallel, so the relative time per separation is not adversely impaired compared with those of other separation techniques. If a relatively large plate number is to be achieved then high pressures are required. Reducing the particle size from an average of 5 μm to 3 μm result in improved efficiency and favorable separation times, but is more demanding in terms of operating

Table 6.3
Characteristic properties of silica gel precoated layers and columns

| Parameter | | Layers | | Column |
|---|---|---|---|---|
| | | High performance | Conventional | High performance |
| Porosity | Total | 0.65-0.70 | 0.65-0.75 | 0.8-0.9 |
| | Interparticle | 0.35-0.45 | 0.35-0.45 | 0.4-0.5 |
| | Intraparticle | 0.28 | 0.28 | 0.4-0.5 |
| Flow resistance parameter | | 875-1500 | 600-1200 | 500-1000 |
| Apparent particle size (μm) | | 5-7 | 8-10 | $d_p$ |
| Minimum plate height (μm) | | 22-25 | 35-45 | 2-3 $d_p$ |
| Optimum velocity (mm/s) | | 0.3-0.5 | 0.2-0.5 | 2 |
| Minimum reduced plate height | | 3.5-4.5 | 3.5-4.5 | 1.5-3 |
| Optimum reduced velocity | | 0.7-1.0 | 0.6-1.2 | 3-5 |
| Knox equation coefficients | | | | |
| Flow anisotropy (A) | | 0.4-0.8 | 1.7-2.8 | 0.5-1.0 |
| Longitudinal diffusion (B) | | 1.2-1.6 | 1.2-2.0 | 1-4 |
| Resistance to mass transfer (C) | | 1.4-2.4 | 0.70-0.85 | 0.05 |
| Separation impedance | | 10,000-20,000 | 11,000-13,000 | 2000-9000 |
| Mean pore diameter (Si 60) (nm) | | 5.9-7.0 | 6.1-7.0 | |

pressure. There are no expected benefits from the use of conventional layers for forced flow separations.

### 6.3.4 Kinetic Properties of Precoated Layers

Layers have a heterogeneous structure consisting of the sorbent particles, held together by added binder, with other possible additives, such as a fluorescence indicator for visualization of UV-absorbing compounds. Typical values for the kinetic properties precoated silica gel layers are summarized in Table 6.3 [11,38,61,62,65,66]. The similar values for the interparticle porosity of columns and layers suggests that their packing density is similar, whereas the significant difference in the total porosity and intraparticle porosity indicates that the intraparticle volume of the layers is substantially smaller. A significant amount of the binder used in stabilizing the layers must be contained within the pores [38]. The mean pore size and pore size distribution of the precoated layers determined by size-exclusion chromatography are in accordance with expected values for the silica gel, tending to confirm that the binder resides largely within the pores and does not specifically block the pore entrances [65]. The difference between the flow resistance parameter of the high performance and conventional layers might arise from the presence of a greater proportion of fine particles in the high performance layer compared with the conventional layer, reflecting the fact that high performance layers are prepared from particles of a smaller average particle size initially. Layers with a narrower particle size distribution but similar average particle size could lead to a further improvement in capillary flow separations on high performance layers. The small difference between apparent average particle sizes for high performance and conventional layers is the origin of the small difference between the minimum plate height for modern layers. Historically, conventional layers afforded poorer separations because they were prepared from sorbents with a significantly wider particle size distribution. Contemporary conventional layers are prepared from particles of smaller average size and narrower particle size distribution compared with a decade ago, reflected in their improved separation performance.

The reasons for the difference in kinetic performance between columns and layers can be deduced from the coefficients of the Knox equation, Table 6.3. The A coefficient is a measure of flow anisotropy within the streaming part of the mobile phase and is related to the packing density and homogeneity of the layer. For high performance layers the A coefficients are smaller than those for typical columns, indicating that the layers are homogeneously packed and have a good packing structure. Conventional layers are not packed as homogeneously, perhaps because of the greater difficulty of preparing layers from sorbents with a broader particle size distribution. The C coefficient is a measure of the resistance to mass transfer between the stationary phase and the streaming portion of the mobile phase. Typical values of C for columns are much smaller than for layers. The large resistance to mass transfer term might arise as a result of restricted diffusion within the porous particles or from partial blocking of the intraparticle channels, owing to the presence of binder. This is probably of little account for capillary flow separations owing to the low prevailing mobile phase velocities but is significant for separations by forced flow. The B coefficient is a measure of the contribution of longitudinal diffusion to the plate height and should be similar for layers and columns, as observed.

### 6.3.5 Zone Broadening in Multiple Development

Unidimensional multiple development provides a complementary approach to forced flow for minimizing zone broadening [61,67-70]. All multiple development techniques employ successive repeated development of the layer in the same direction with the removal of mobile phase between developments. Approaches differ in the changes made, e.g. mobile phase composition and solvent-front migration distance, between consecutive development steps; the total number of consecutive development employed can also be varied. Capillary forces are responsible for migration of the mobile phase but a zone-focusing mechanism is used to counteract the normal zone broadening that occurs in each successive development. Each time the solvent front traverses the stationary sample zone, the zone is compressed in the direction of development. The compression occurs because the mobile phase contacts the bottom edge of the zone first; here the sample molecules start to move forward before those molecules still ahead of the solvent front. When the solvent front has reached beyond the zone, the focused zone migrates and is subject to the normal zone broadening mechanisms. Both theory and experiment indicate that beyond a minimum number of development steps zone widths converge to a constant value that is roughly independent of migration distance, Figure 6.5 [71]. The position of a zone in multiple chromatography (constant development length) is given by

$$R_{F,P} = 1 - (1 - R_F)^p \qquad (6.7)$$

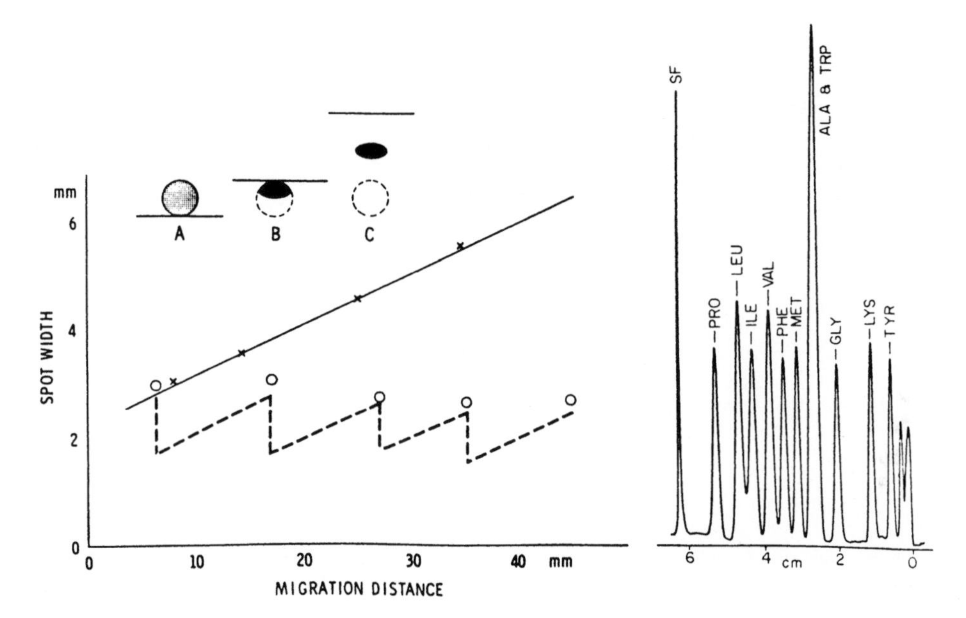


Figure 6.5. Illustration of the zone refocusing mechanism in multiple development (left) and its application to the separation of a mixture of phenylthiohydantoin-amino acid derivatives (right). The broken line represents the change in zone size due to the expansion and contraction stages in multiple development and the solid line depicts the expected zone width for a zone migrating the same distance in a single development. (From ref. [71]; ©Research Institute for Medicinal Plants).

and in incremental multiple development (each successive development distance is increased by a constant amount) by

$$R_{F,P} = 1 - (1 - R_F)[\{1 - (1 - R_F)^p\} / pR_F] \qquad (6.8)$$

where $R_{F,P}$ is the apparent $R_F$ value after p successive developments and $R_F$ is the $R_F$ value in the first development. Equations used to predict zone widths are more complicated and better handled by computer [69,70]. The theory developed so far is for isocratic mobile phases and is not applicable to solvent gradients, although the same general phenomenon of a roughly constant zone width is observed in optimized separations by automated multiple development. For difficult separations by capillary flow, multiple development is the general strategy used to increase the zone capacity.

### 6.3.6 Resolution and Zone Capacity

The resolution, $R_S$, between two separated zones in thin-layer chromatography is defined as follows

$$R_S = 2(Z_{X2} - Z_{X1}) / (w_{b1} + w_{b2}) \qquad (6.9)$$

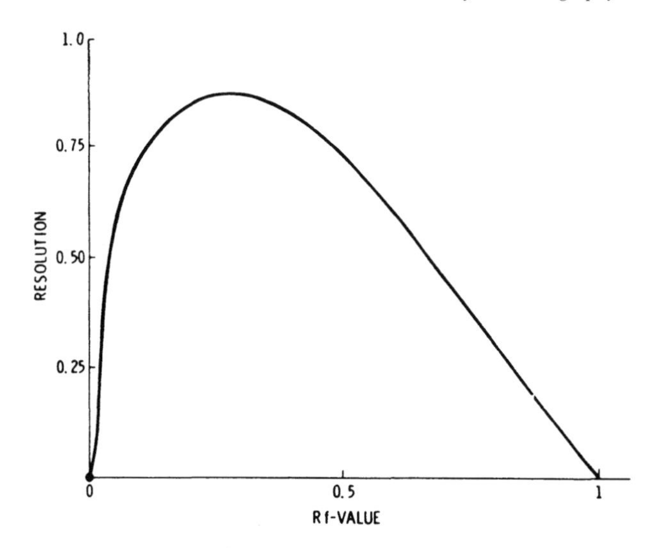


Figure 6.6. Variation of the resolution of two closely migrating zones as a function of the $R_F$ value of the faster moving zone.

where $Z_X$ is the zone center migration distance, $w_b$ the width of the zone at its base, and the subscripts 1 and 2 refer to the individual zones with the larger number corresponding to the zone with the higher $R_F$ value [72]. To optimize resolution, it is necessary to relate resolution to the experimental parameters, such as the layer efficiency, the ratio of the equilibrium constants for the separation process, and the position of the zones in the chromatogram. The thin-layer analog of classic resolution equation for column chromatography is given by

$$R_S = (\sqrt{N} / 4)\,(R_{F2} - R_{F1}) / R_{F2} \tag{6.10}$$

However, in thin-layer chromatography neither N nor the retardation factors are independent of the ratio of the equilibrium constants for the separation process or the properties of the layer. The value of N is strongly dependent on $R_F$ for capillary flow. As a crude approximation N can be replaced by $N_1 R_{F2}$ where $N_1$ is the plate number the zones would have passed over if they migrated to the solvent front position. Substituting for N in equation (6.10) and rearranging gives

$$R_S = [\sqrt{(N_1 R_{F2})} / 4]\,[(k_1 / k_2) - 1]\,[1 - R_{F2}] \tag{6.11}$$

where k is the retention factor. Equation (6.11) is adequate for a qualitative description of the effect of layer quality, selectivity, and zone position on resolution for a single development with capillary flow, Figure 6.6. Resolution increases with efficiency in a manner that depends linearly on the $R_F$ value. Relatively small changes in selectivity have a large impact on the ease of obtaining a separation, since the total plate number available is not very large. Separations by thin-layer chromatography are easy when $(R_{F2} - R_{F1}) > 0.1$, and very difficult or impossible for $(R_{F2} - R_{F1}) < 0.05$, in the region of the optimum $R_F$ value for the separation. The effect of zone location on resolution shows the opposite behavior to layer quality. At large values of $R_{F2}$, the term $(1 - R_{F2})$ is small and resolution falls to zero at $R_{F2} = 1$. Differentiation of Eq. (6.11) indicates that the maximum resolution of two difficult to separate zones will occur at a $R_F$ value of about 0.3. From Figure 6.6, it can be seen that resolution does not change significantly for $R_F$ values between 0.2 and 0.5; and within this range, the resolution is greater than 92% of the maximum value (75% between $R_F = 0.1$ and 0.6).

Resolution in forced flow thin-layer chromatography is not restricted by the same factors that apply to capillary flow. Resolution increases almost linearly with the solvent-front migration distance and is highest for separations at the optimum mobile phase velocity. Resolution has no theoretical limit for forced flow; the upper bounds are established by practical constraints (plate length, separation time and inlet pressure).

The potential of a chromatographic system to provide a separation can be estimated from its zone capacity, also referred to as the separation number (SN) or spot capacity in thin-layer chromatography. The separation number envisages the chromatogram as being similar to a string of beads, each bead touching its neighbor, with no unoccupied space between the beads. In practice, it leads to an inflated estimate of the real separation capacity, since real chromatograms do not consist of equally spaced peaks. It does provide, however, a plausible guide for comparison of different thin-layer separation systems. In addition, it provides an indication of the possibility of separating a given mixture, for unless the separation number exceeds the number of sample components by a significant amount, the separation will be difficult to impossible to achieve.

The separation number is defined as the number of spots completely separated with $R_S = 1$ between $R_F = 0$ and $R_F = 1$. Most methods of calculation, however, make approximations for either the form of the chromatogram, such as a geometric increase in zone size throughout the chromatogram [7,32,73], or attempt to estimate the increase in zone size by a theoretical model [74,75]. No approach is completely satisfactory, and only the simplest case is considered here. The separation number is calculated from Eq. (6.12), or more simply Eq. (6.13), with $b_0$ and $b_1$ as defined in Figure 6.2.

$$SN = \log (b_0 / b_1) / [\log (1 - b_1 + b_0) / (1 + b_1 - b_0)] \quad (6.12)$$

$$SN = [(Z_f - Z_o) / (b_0 + b_1)] - 1 \quad (6.13)$$

Equation (6.13) provides values that are about 25% smaller that those calculated by Eq. (6.12). Some typical results from theory or determined by experiment are summarized in Table 6.4. Results from theory are probably too high and represent an upper limit. Experiment indicates a zone capacity of about 12–14 for a single development with capillary flow. This rises to about 30 – 40 for forced flow. Automated multiple development with capillary flow provides a similar zone capacity to forced flow. Two-dimensional thin-layer chromatography employing different retention

Table 6.4
Experimental or theoretical zone capacity for different conditions in thin-layer chromatography

| Separation mode | Dimension | Zone capacity |
|---|---|---|
| *(i) Predictions from theory* | | |
| Capillary flow | 1 | < 25 |
| Forced flow | 1 | < 80 (up to 150 depending on pressure limit) |
| Capillary flow | 2 | < 400 |
| Forced flow | 2 | Several thousand |
| *(ii) Experimental observations* | | |
| Capillary flow | 1 | 12 - 14 |
| Forced flow | 1 | 30 - 40 |
| Capillary flow (AMD) | 1 | 30 - 40 |
| Capillary flow | 2 | ≈ 100 |
| *(iii) Predictions based on (ii)* | | |
| Forced flow | 2 | ≈ 1500 |
| Capillary flow (AMD) | 2 | ≈ 1500 |

mechanisms for each orthogonal separation provides a significantly larger zone capacity than any of the unidimensional development methods. This accounts for the continuing interest in this technique (see section 6.6.4). In theory, the zone capacity becomes very large for two-dimensional thin-layer chromatography if forced flow or multiple development is used for the orthogonal separations, but this has not been demonstrated by experiment [1].

## 6.4 STATIONARY PHASES

Thin-layer plates can be prepared in the laboratory by standardized procedures [5,6], although the exacting experimental conditions required for the preparation of layers with reproducible separation properties are more easily obtained in a manufacturing setting. Today, most laboratories use commercially available precoated plates. Precoated plates for high performance, conventional and preparative thin-layer chromatography are available in thicknesses from 0.1 to 2.0 mm on either glass, aluminum or plastic backing sheets [76]. Most plates also contain a binder such as poly(acrylic acid) and its salts, gypsum or starch in amounts from 0.1 to 10% (w/w), to impart the desired mechanical strength, durability and abrasion resistance to the layer. A UV-indicator, such as manganese-activated zinc silicate of a similar particle size to the sorbent, may be added to the layer for visual evaluation of sample zones by fluorescence quenching. Thin-layer plates with a binary layer of two different, separated sorbents, forming a narrow interface parallel to one edge, are available for two-dimensional separations. If one of the layers is a type of silica with very weak retention properties, it can be used as a concentrating zone, to aid sample application [77]. Recent commercial

Table 6.5
Typical properties of inorganic oxide adsorbents used for thin-layer chromatography

| Property | Silica gel | Alumina | Kieselguhr |
|---|---|---|---|
| Specific surface area ($m^2$ / g) | 200-800 | 50-350 | 1-4 |
| Specific pore volume (ml / g) | 0.5-2 | 0.1-0.4 | 1-3 |
| Average pore diameter (nm) | 4-20 | 2-35 | $10^3$-$10^4$ |
| Concentration of active sites (μmol / $m^2$) | 8 | 13 | |

Silica gel high performance layers

| | Merck Si 60 | Merck Si 50000 | Whatman HP-K |
|---|---|---|---|
| Specific surface area ($m^2$ / g) | 550 | 0.5 | 300 |
| Specific pore volume (ml / g) | 0.82 | ≈0.63 | 0.70 |
| Average pore diameter (nm) | 6 | 5000 | 8 |
| Average particle size (μm) | 5 | 5 | 5 |
| Layer thickness (mm) | 0.2 | 0.2 | 0.2 |
| Apparent pH (10 % aqueous suspension) | 7.0 | | 7.0-7.2 |

developments include high performance layers prepared from spherical silica particles and mixed silica and magnesium tungstate layers for improved sensitivity with wider spectral windows for *in situ* diffuse reflectance FTIR measurements [78]. Binder-free, ultra-thin monolithic silica layers for fast separations prepared by sol-gel technology have been prepared [79]. Glass rods with a sintered adsorbent coating are used for thin-layer chromatography, primarily in conjunction with a scanning flame ionization detector (section 6.9.5) [80,81].

### 6.4.1 Inorganic Oxide Layers

Inorganic oxide adsorbents commonly used in thin-layer chromatography include silica gel, alumina, kieselguhr, and Florisil [9,82]. Of these, silica gel is by far the most important. The chromatographic properties of inorganic oxides depends on their surface chemistry and composition, specific surface area, specific pore volume, and average pore diameter, Table 6.5. Retention on silica gel depends on the number, type and spatial location of analyte functional groups. Surface silanol groups are the dominant sites for analyte interactions. Since these are present at roughly the same concentration for all silica surfaces, general retention largely reflects differences in the specific surface area. More specific differences in the type and distribution of silanol groups are responsible for selectivity differences [83]. Consequently, it is not always possible to reproduce the same separation on different silica gel layers. The influence of analyte functional group type and spatial position on retention is illustrated in Figure 6.7 for the separation of ethynyl steroids, which are the active ingredients in oral contraceptives [1,84]. The steroids with phenolic groups are the most strongly retained, followed by hydroxyl groups, and ketone and ester groups. Subtle separation differences are also seen due to steric hindrance at a functional group and differences in ring conformations. Silica gels of extremely low surface area (e.g. Si 50000) are used for the separation of polar

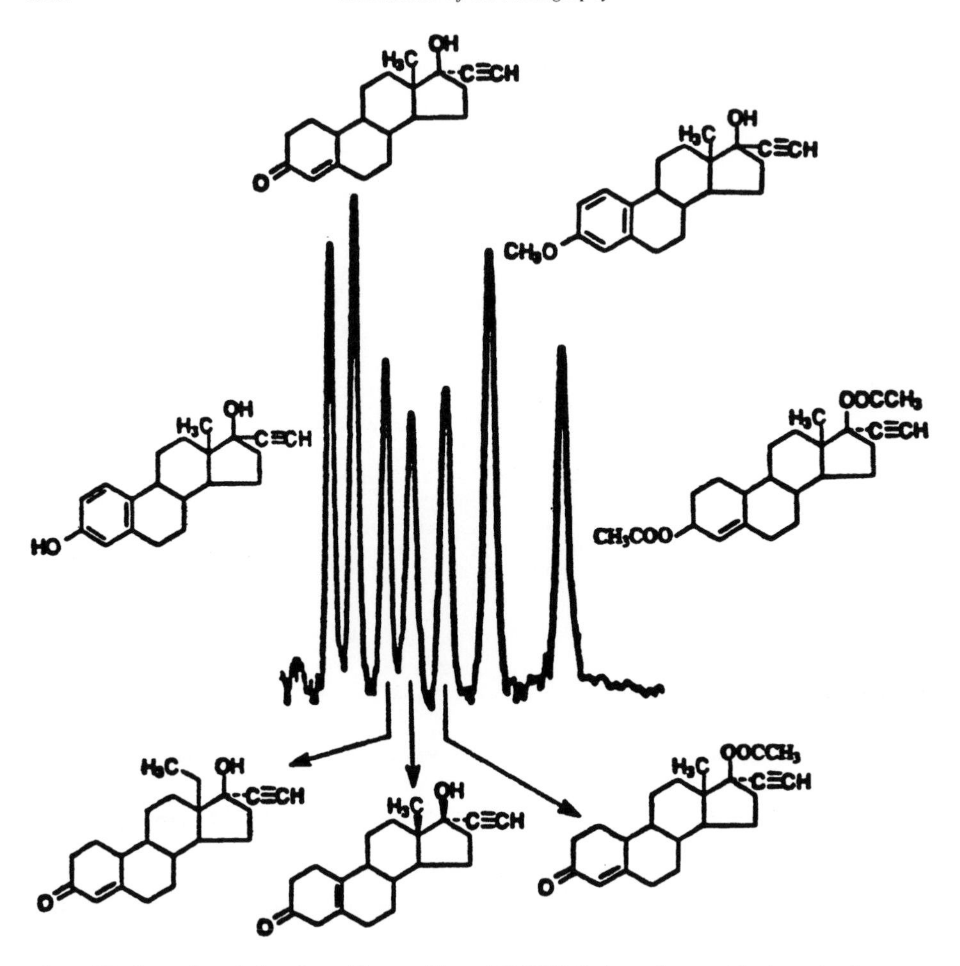

Figure 6.7. Separation of ethynyl steroids on a silica gel 60 HPTLC plate using two 15-minute developments in hexane-chloroform-carbon tetrachloride-ethanol (7:18:22:1) as mobile phase. (From ref. [1]; ©Elsevier).

compounds (e.g. carbohydrates, nucleic acid derivatives, phosphates and sulfonates, etc.) that are not easily separated on high-activity silica gel layers [85]. This type of silica gel is the same as that used for the preparation of concentrating zones on bilayer plates. With the exception of Si 50000 the range of adsorbents with different pore sizes is not great. Most materials have an average pore size of 6 nm designed for the separation of small molecules (MW < 750). The thickness of high performance layers is typically 0.2 mm. Reducing the layer thickness to 0.1 mm increases the separation speed (mobile phase velocity constant increases about 1.1-2.5 fold) but has little influence on the zone capacity [86]. The thinner layers afford an increase in detection limits by scanning densitometry of about 1.1 to 1.5 fold. These small gains in speed and limits of detection are offset by a loss in ease of use. Impregnating silica gel layers with a solution of a complexing agent is a convenient method for modifying the selectivity of the layer for specific applications [5,6,87-89]. A few common examples include silver nitrate for the separation of saturated and unsaturated compounds, boric acid to differentiate between isomers with vicinal hydrogen-bonding functional groups, and caffeine or picric acids for the separation of polycyclic aromatic compounds.

The separation properties of alumina are complementary to those of silica gel [82,90,91]. The alumina surface is more complex containing hydroxyl groups, aluminum cations, and oxide anions. Its apparent pH and hydration level significantly influences the separation properties of alumina. Alumina is available with various adjusted pH ranges, corresponding to basic (pH $\approx$ 9-10), neutral (pH $\approx$ 7-8) and acidic (pH $\approx$ 4-4.5). The separation mechanism for alumina is not as well understood as silica gel. Kieselguhr (a naturally occurring impure form of silica gel) and Florisil are not widely used in contemporary practice [82,91].

### 6.4.2 Chemically Bonded Layers

Chemically bonded layers are prepared by reacting silanol groups on the silica gel surface with various organosilane reagents [9,82,92-94]. Silica-based, chemically bonded layers with dimethyl-, diphenyl-, ethyl-, octyl-, octadecyl-, 3-aminopropyl-, and 3-cyanopropylsiloxane-bonded groups and a siloxane-bonded spacer bonded propanediol group are commercially available. The selection of silica gel substrate and method of preparation result in nominally similar generic products with different separation properties, Table 6.6. Reversed-phase layers from Merck are prepared from dichloroalkylmethylsilanes with methyl- (RP-2), octyl- (RP-8), and octadecyl- (RP-18) alkyl groups. The organosilane reagent is bonded to the surface by a combination of one or two bonds. The 3-aminopropylsiloxane-bonded layer is prepared with 3-aminopropyltriethoxysilane and has a much higher surface coverage than the 3-cyanopropylsiloxane-boded layer prepared from 3-cyanopropyldichloromethylsilane. The 3-cyanopropylsiloxane group is bound almost entirely to the silica substrate by bidentate linkages. On the other hand, the 3-aminopropylsiloxane group is bound to the silica substrate by a combination of bidentate and tridentate linkages. In the latter case, this is most likely the result of crosslinking of the reagent during the surface reaction. The two Macherey-Nagel octadecylsiloxane-bonded layers are of the polymeric type, prepared from a trifunctional reagent. They differ primarily in the extent of surface bonding. The Whatman products are prepared by different chemistries. The dimethylethylsiloxane-bonded and dimethyloctylsiloxane-bonded layers are prepared from monofunctional organosilane reagents. The octadecylsiloxane-bonded layer is prepared from a trifunctional organosilane reagent and has a crosslinked, polymeric structure. In addition, the octylsiloxane-, octadecysiloxane- and diphenylsiloxane-bonded layers are endcapped, but even so, they posses a relatively high concentration of silanol groups.

Reversed-phase alkylsiloxane-bonded layers with a high level of surface bonding are incompatible with mobile phases containing a significant amount of water. In this

Table 6.6
Characteristic properties of precoated chemically bonded layers

| Manufacturer | Derivatizing reagent | Percent silanol groups reacted | Carbon loading (%) | Average particle size (μm) |
|---|---|---|---|---|
| Merck | | | | |
| RP-2 | Bifunctional | 50 | | 5-7 |
| RP-8 W | Bifunctional | 25 | 8.9 | 11-13 |
| RP-8 | Bifunctional | 37 | | 5-7 |
| RP-18 W | Bifunctional | 22 | 15.4 | 11-13 |
| RP-18 | Bifunctional | 35 | | 5-7 |
| 3-Aminopropyl | Trifunctional | 50 | 5.8 | 5-7 |
| 3-Cyanopropyl | Bifunctional | 27 | | 5-7 |
| Whatman | | | | |
| KC-2 | Monofunctional | | 4.5 | 10-14, or 20 |
| KC-8 | Monofunctional | | 8.5 | 10-14 |
| KC-18 | Trifunctional | 16 | 12.5 | 10-14 |
| Diphenyl | Difunctional | | 8.5 | 10-14 |
| Macherey-Nagel | | | | |
| Sil C18-100 | Trifunctional | 45 | | 5-10 |
| Sil C18-50 | Trifunctional | 30 | | 5-10 |

case, the hydrophobic repulsive forces are stronger than the capillary forces moving the solvent through the layer (this does not apply to forced flow separations). For mobile phases containing more than about 40% (v/v) water, the mobile phase velocity is very slow and/or the precoated layers swell and flake off from the support. This problem is overcome by using a slightly larger particle size for the silica substrate, a reproducible, although lower level of silanization than the maximum value, and a different binder to enhance its water stability. These layers are referred to as water wettable and are used for all types of reversed-phase separations, while layers with a high level of silanization are used for normal-phase separations with predominantly non-aqueous mobile phases. For reversed-phase separations, the range of mobile phase velocities depends in a complex manner on the mobile phase composition, Figure 6.8 [93-95]. Contributing to the shape of the curves are properties of the binder as well as the mobile phase (e.g. surface tension/viscosity ratio and possibly the contact angle). Polar chemically bonded phases are compatible with water in all proportions and are suitable for use in both normal- and reversed-phase chromatography. In addition, the mobile phase velocity varies less with composition. Hydrolysis of bonded phases by partially aqueous, extreme pH, mobile phases was shown to affect separations using multiple development [96].

Chemically bonded layers are used essentially for separations that cannot be performed on silica gel, Table 6.7. Compounds of extreme polarity and ionic compounds are difficult to separate on silica gel because of strong retention and limited selectivity. Chemically bonded phases provide complementary properties to silica gel for the separation of compounds of intermediate polarity. In addition, 3-aminopropylsiloxane-bonded layers are used as weak ion-exchangers for the separation of anions with acid-

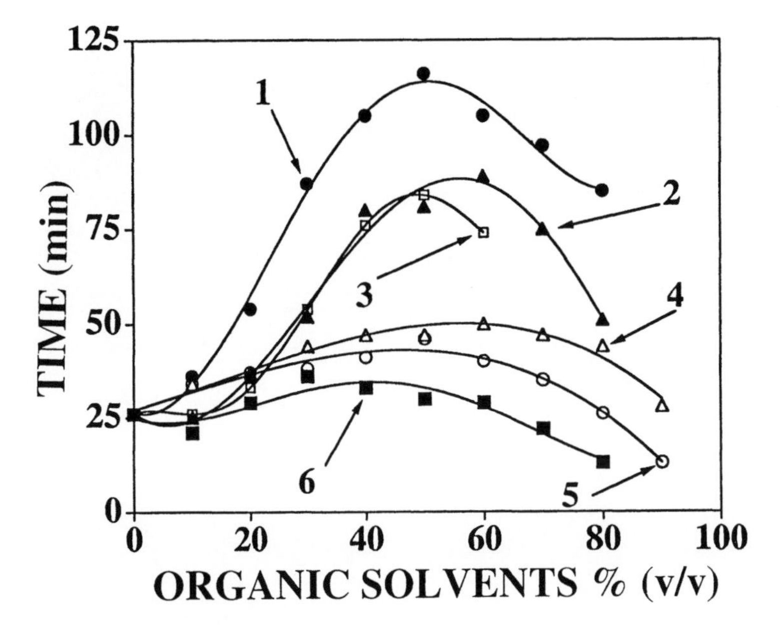


Figure 6.8. Plot of the time required for the solvent front to migrate 5 cm as a function of mobile phase composition for 2-propanol (1), N,N-dimethylformamide (2), 2,2,2-trifluoroethanol (3), methanol (4), acetone (5) and acetonitrile (6) in water on a Merck HPTLC RP-18 WF254s layer. (From ref. [95]; ©Research Institute for Medicinal Plants).

buffered mobile phases [12,97,98]. A useful property of 3-aminopropylsiloxane-bonded layers is the ease of converting different substances that may require derivatization for detection (e.g. carbohydrates, amino acids, and catecholamines) into fluorescent derivatives simply by heating while adsorbed on the layer [99].

### 6.4.3 Other Materials

Native and microcrystalline cellulose precoated plates are used in the life sciences for the separation of polar compounds (e.g. carbohydrates, carboxylic acids, amino acids, nucleic acid derivatives, phosphates, etc) [85]. These layers are unsuitable for the separation of compounds of low water solubility unless first modified, for example, by acetylation. Several chemically bonded layers have been described for the separation of enantiomers (section 10.5.3). Polyamide and polymeric ion-exchange resins are available in a low performance grade only for the preparation of laboratory-made layers [82]. Polyamide layers are useful for the reversed-phase separation and qualitative analysis of phenols, amino acid derivatives, heterocyclic nitrogen compounds, and carboxylic and sulfonic acids. Ion-exchange layers prepared from poly(ethyleneimine), functionalized poly(styrene-divinylbenzene) and diethylaminoethyl cellulose resins and powders and are used primarily for the separation of inorganic ions and biopolymers.

Table 6.7
Retention properties of silica-based chemically bonded layers

| Type of modification | Functional group | Application |
|---|---|---|
| Alkylsiloxane | Si-$CH_3$<br>Si-$C_2H_5$<br>Si-$C_8H_{17}$<br>Si-$C_{18}H_{37}$ | • For reversed phase separations generally but not exclusively<br>• Separation of water soluble polar organic compounds (RPC)<br>• Weak acids and bases after ion suppression (RPC)<br>• Strong acids and bases by ion-pair mechanism (RPC)<br>• Homologs and Oligomers (RPC)<br>• Hydrocarbon-like and polycyclic aromatic compounds (RPC & NPC) |
| Phenylsiloxane | Si-$C_6H_5$ | • No specific applications that cannot also be preformed on alkylsiloxane-bonded layers |
| Cyanopropylsiloxane | Si-$(CH_2)_3CN$ | • Useful for both RPC and NPC<br>• In NPC it exhibits properties similar to a low capacity silica gel.<br>• In RPC it exhibits properties similar to short-chain alkylsiloxane-bonded layers (it has no selectivity for dipole-type interactions) |
| Aminopropylsiloxane | Si-$(CH_2)_3NH_2$ | • Used mainly in NPC & IEC. Limited retention in RPC<br>• Selectively retains compounds by hydrogen-bond interactions in NPC. Separations unlike those obtained on silica gel.<br>• Functions as a weak anion exchanger in acidic mobile phases (IEC) |
| Spacer bonded propanediol | Si-$(CH_2)_3OCH_2CH(OH)CH_2OH$ | • Used in NPC and RPC but more useful for NPC because of low retention in RPC.<br>• Reasonable retention of polar compounds by hydrogen bond and dipole-type interactions in NPC. More hydrogen-bond acidic and less hydrogen-bond basic then aminopropylsiloxane-bonded layers in NPC. More retentive than aminopropylsiloxane-bonded layers in RPC.<br>• Similar retention to short chain alkylsiloxane-bonded layers but different selectivity for hydrogen- bonding compounds |

Several types of crosslinked gels [e.g. dextrans (Sephadex), poly(acrylamide) (Bio-Gel P), and poly(acryloylmorpholine) (Enzacryl)] are used for the separation of biopolymers by thin-layer chromatography. These materials must be swollen in an appropriate buffer/solvent system before spreading as a layer. Laboratory-made layers require modified equipment for their use and afford only modest performance characteristics.

### 6.4.4 Layer Pretreatments

Before chromatography, it is common practice to prepare the layers for use by any or all of the following pretreatments: washing, activation, conditioning, and equilibration. In addition, layers may be cut to appropriate sizes using scissors for plastic- or aluminum-backed plates and diamond or carbide glass cutting tools for glass-backed layers [100]. Precoated plates are invariably contaminated, or quickly become so, because of residual contaminants from the manufacturing process, contact with packaging materials, and adsorption of materials from the atmosphere. Contamination may result in irregular and drifting densitometric baselines, ghost peaks in the chromatogram (recognized from a chromatogram obtained by scanning the layer between sample tracks), and reduced sensitivity for postchromatographic derivatization [101]. These problems are easily remedied by predevelopment and/or immersion of the layer in a polar solvent such as methanol or 2-propanol [100-102]. Predevelopment with a strong solvent or the intended mobile phase for the separation, often results in a pile up of contaminants in a region close to the top of the plate. This can result in an upwardly sloping baseline in scanning densitometry. A single or double immersion in a polar solvent for about 5 minutes is generally superior to predevelopment for removing contaminants. For trace analysis, both sequential immersion and predevelopment may be required for acceptable results [102,103].

The absolute $R_F$ value and its reproducibility on inorganic oxide layers depends on the layer activity. This is controlled by the adsorption of reagents, most notably water, through the gas phase [100]. Physically adsorbed water is removed from silica gel layers by heating at about 120°C for 30 minutes. Afterwards the plates are stored over a drying agent in a grease-free desiccator. Heat activation is not usually required for chemically bonded layers. Deactivation of silica gel layers by exposure to the atmosphere is extremely rapid. In modern environment-controlled laboratories, layers achieve a consistent level of activity almost instantaneously, that should provide sufficient reproducibility for most separations. Indeed, brief exposure of activated layers to the laboratory atmosphere may render the activation process a waste of effort.

Inorganic oxide layers can be adjusted to a defined activity level by exposure to a defined gas phase in an enclosed chamber. This is best performed after sample application in a developing chamber that allows both conditioning and development of the layer in the same chamber (e.g. a twin trough chamber). Alternatively, separate conditioning and development chambers can be used. Atmospheres of different constant relative humidity can be obtained by exposure to the vapor phase in equilibrium with solutions of concentrated sulfuric acid or saturated solutions of various salts [100]. In the same way, acid or base deactivation is carried out by exposure to concentrated ammonia or hydrochloric acid fumes.

## 6.5 SAMPLE APPLICATION

Samples are applied to thin-layer plates as spots or bands to conform to the demands of minimum size and a homogeneous distribution of sample within the starting zone [8,9,104,105]. For high performance layers, starting spot sizes of 1 – 2 mm are desirable, corresponding to a sample volume of 100 to 200 nl if applied by a dosimeter. For conventional layers, sample volumes 5 to 10-fold higher are acceptable. Favorable properties for sample solvents are summarized in Table 6.8. Sample application by hand-held devices is unsuitable for use with scanning densitometry, which requires that the starting position of each spot be accurately known. This is easily achieved with

Table 6.8
Solution requirements for sample application

| Property | Requirements |
|---|---|
| Sample solvent | • Good solvent for the sample to promote quantitative transfer from the sample application device to the layer<br>• Low viscosity and sufficiently volatile to be easily evaporated from the layer<br>• Wet the sorbent layer to provide adequate penetration of the layer by the sample (a potential problem for alkylsiloxane-bonded layers and aqueous sample solutions)<br>• Weak chromatographic solvent to minimize predevelopment during sample application (ideally if used as a chromatographic solvent the least retained sample component should have an $R_F$ value < 0.1) |
| Aqueous solutions | • Dilute if possible with a water-miscible solvent that forms a lower boiling azeotrope<br>• Apply in small increments or if spray-on techniques are used with a slow application rate<br>• Use layers with a concentrating zone and refocus the sample prior to development |
| Viscous solutions | • Dilute if possible with a volatile solvent of low viscosity<br>• Use a direct transfer technique (e.g. contact spotter) |
| Suspensions | • Filter before attempting sample application<br>• Otherwise use layers with a concentrating zone and an extraction solvent that mobilizes the components of interest for refocusing |

mechanical devices operating to a precise grid mechanism. In addition, the sample must be applied to the layer without damaging the surface, something that is near impossible to achieve with manual sample application devices. When using dosimeters, it is important that samples and standards are applied to the layer in identical volumes and in the same solvent. This ensures identical starting zones are formed. Calibration curves should not be prepared by applying variable volumes from a single standard solution, since this is likely to result in poor linearity and non-zero intercepts. This restriction does not apply to spray-on devices.

### 6.5.1 Application Devices

Sample application devices for thin-layer chromatography encompass a wide range of sophistication and automation. The most popular devices for quantitative thin-layer chromatography use the spray-on technique. A nitrogen atomizer sprays the sample from a syringe or capillary onto the layer surface as narrow, homogeneous bands. Linear or zigzag motion of either the plate on a motorized stage or the atomizer head allows bands of any length between 0 (spots) and the maximum transit length of the spray head (or plate). Typical band lengths are 0.5 or 1.0 cm with longer bands used mainly for preparative-scale separations. The rate of sample deposition is also adjustable to accommodate sample solutions of different volatility and viscosity. Band application allows the sample band to be made longer than the slit length of the light source, minimizing quantification errors due to positioning of the sample within the light beam in scanning densitometry. Also, different volumes of a single standard can be applied for calibration and the standard addition method of quantification is facilitated by overspraying the sample already applied to the layer with a solution of the standard.

Several fully automated spray-on sample applicators are available. In one device, a motor driven syringe is used to suck up sample volumes of 0.1 to 50 μl, which are then deposited as spots or bands on the layer [104]. The syringe feeds a stainless-steel capillary connected to a capillary atomizer. The applicator can be programmed to select samples from a rack of vials and deposit fixed volumes of the sample, at a controlled rate, to selected positions on the layer. The applicator automatically rinses itself between applications and can spot or band a whole plate with different samples and standards without operator intervention. A number of multi-sample applicators for the simultaneous transfer and deposition of several samples at the same time have been described [106-108].

Glass microcapillaries and fixed-volume dosimeters are also used for sample application. These require less sophisticated instruments. Fixed-volume dosimeters with 100 or 200 nl volumes for use with high performance layers are fabricated from platinum-iridium capillaries sealed into a glass support capillary of larger bore [7]. The metal capillary tip is polished to provide a smooth flat surface of small area (ca. 0.05 $mm^2$), which is brought into contact with the layer by a mechanical device to discharge its volume. Mechanical application of the sample is possible by attaching a metal collar to the glass support capillary. This allows the dosimeter to be held by a magnet and lowered to the plate surface under controlled conditions. A spring mechanism allows the applicator head to be lowered and lifted from the plate surface while the frictional force holding the dosimeter to the applicator head controls the force with which the dosimeter engages the layer. A click-stop grid mechanism allows the even spacing of sample spots on the layer, and provides a frame of reference for sample location during scanning densitometry.

Layers with a concentrating zone simplify some aspects of sample application [77]. Microliter volumes can be applied either as spots or bands to any position on the concentrating zone. Alternatively, the entire zone can be immersed in a dilute solution of the sample. During development, the sample migrates out of the concentrating zone and is focused to a narrow band at the interface between the concentrating zone and separation layer. However, since the distribution of the sample within the band may be uneven, quantitative measurements by scanning densitometry are often poor. Thin-layer plates with concentrating zones are mainly used for qualitative applications, for the application of large sample volumes, and for the application of crude samples (e.g. biological fluids).

## 6.6 MULTIMODAL (COUPLED COLUMN-LAYER) SYSTEMS

The specific reasons for coupling column separations with thin-layer chromatography are to increase peak capacity and to take advantage of layer attributes summarized in Table 6.1 (section 6.2). The thin-layer plate functions as a separation and storage device retaining information from the column and thin-layer separations in an immobilized format. This allows sample components to be investigated free of time constraints, which is advantageous for: biomonitoring (section 6.9.6); for samples that require derivatization for detection; for sequential evaluation using different detection principles; to preserve and transport the separation to different locations for evaluation; and for applications employing solid-phase spectroscopic identification techniques (section 9.3.1.4).

Coupling gas chromatography to thin-layer chromatography (GC-TLC) is straightforward but has not been widely used since the late 1960s when several interfaces were described [61]. Many of the problems for which these instruments were used are solved by gas chromatography-mass spectrometry today, and contemporary interest in GC-TLC has declined. In the late 1970s an apparatus for supercritical fluid extraction with deposition of the fluid extract onto a moving thin-layer plate was described [109,110]. A suitable interface for direct coupling of supercritical fluid chromatography to thin-layer chromatography (SFC-TLC) was described more recently [111,112]. Decompression of the supercritical fluid at a capillary orifice occurs with rapid cooling, favoring the deposition process without disturbing the conversion of the fluid to a gas. Efficient transfer of sample to the layer requires solvent addition, to minimize the loss of high-speed sample particles. Wet particle deposition being much more efficient than dry particle deposition. The coupling of capillary electrophoresis to thin-layer chromatography (CE-TLC) has been described, but no real applications developed [113]. Current interest is largely in the coupling of column liquid chromatography to thin-layer chromatography (LC-TLC), which has reached a reasonable level of maturity and commercialization.

### 6.6.1 Liquid Chromatography-Thin-Layer Chromatography (LC-TLC)

The most general interface for coupling column liquid chromatography to thin-layer chromatography (LC-TLC) is based on different modifications to the spray-jet applicator (section 6.5.1) [61,114-120]. At flow rates typical for packed capillary columns (5–100 μl/min) the total mobile phase can be applied to the layer. A splitter in the transfer line to the spray-jet applicator is required to accommodate higher flow rates from wider bore columns. The column eluent is nebulized by mixing with (heated) nitrogen gas and sprayed as an aerosol onto the layer. The spray head is moved horizontally on one line within a defined bandwidth or, better, is made to deposit the spray over a defined rectangular area (e.g. 8 x 6 mm) to promote effective solvent evaporation. In the latter case, the zone is focused by a short development with a strong solvent before separation.

In the profiling mode, the whole column chromatogram is divided into volume fractions sequentially transferred to the layer and deposited as a series of bands that are subsequently developed in parallel. Each track (band) is scanned individually revealing an immense amount of information about the sample composition. In the target compound mode, fractions identified by the column detector, or from elution windows established by marker compounds, are transferred to the layer and stored there. When all the available space is occupied by column fractions from a single or several samples, the layer is developed and evaluated. The main limitations of the spray-jet interface are its restricted flow capability and inability to handle involatile ionic additives, such as buffers and ion-pair reagents. These problems can be overcome by combining an automated solid-phase extraction module with the spray-jet applicator to concentrate column fractions and exchange the solvent for layer deposition [121].

## 6.7 DEVELOPMENT TECHNIQUES

Development in thin-layer chromatography is the process by which the mobile phase moves through the layer, thereby inducing differential migration of the sample components. The principal techniques are linear, circular and anticircular with the mobile phase velocity controlled by capillary forces or external pressure. The application of any of these techniques can be extended using continuous or multiple development.

### 6.7.1 Linear and Radial Development

For linear development, samples are applied along one edge of the plate and separated in the direction of the opposite edge. Viewed in the direction of development, the chromatogram consists of a series of compact symmetrical spots of increasing diameter, or if samples are applied as bands, as rectangular zones of increasing width.

If the position of sample application and mobile phase entry point are at the center of the layer with the flow of mobile phase towards the periphery, then this mode of development is called circular chromatography [7]. Samples can be injected in the mobile phase, in which case they are separated as a series of concentric rings. Otherwise, samples are applied as a cluster of spots in a radial pattern around the solvent entry position. After development, spots near the origin remain symmetrical and compact while those near the solvent front are compressed in the direction of development and elongated at right angles to this direction [110].

For anticircular development, the sample is applied to the layer along the circumference of an outer circle and developed towards the center [7,123]. Spots near the origin remain compact while those close to the solvent front are elongated in the direction of migration, but of nearly constant width when viewed at right angles to the direction of development. Elongation of the sample spots at high $R_F$ is unavoidable. It arises as a consequence of the lateral compression induced by flow of the mobile phase through a continuously decreasing layer area. The unique features of anticircular development are its high speed and high sample throughput. The mobile phase velocity remains approximately constant under capillary flow conditions because of the equivalence between the quadratic decrease in mobile phase velocity and a similar reduction in the layer area to be wetted.

Both circular and anticircular chromatograms require a scanning densitometer capable of radial or peripheral scanning, an option available with some instruments. Circular development is used in rotational planar chromatography and is popular in preparative-scale chromatography because it allows separated samples to be continuously collected as they elute from the layer. Anticircular development is little used in contemporary practice and requires a specially designed developing chamber. Linear development is "the normal" and most widely used method.

### 6.7.2 Continuous Development

For continuous development, the mobile phase is allowed to traverse the layer under the influence of capillary forces until it reaches a predetermined position on the plate, where it is continuously evaporated. Evaporation of the mobile phase usually occurs at the plate atmospheric boundary by either natural or forced evaporation. The movement of the mobile phase to the air boundary occurs by capillary flow, but once it reaches the boundary, additional forces are applied by evaporation of the solvent. Eventually a steady state (constant velocity) is established, where the mass of solvent evaporating at the boundary is equivalent to the amount of new solvent entering the layer. Sandwich-type chambers for continuous development were reviewed by Soczewinski [124]. Perry [125] has outlined the use of the short-bed continuous development chamber for optimized continuous development with variable selection of the plate length, and Nurok [126] has proposed a theoretical model to optimize experimental conditions. Continuous development is used primarily to separate simple mixtures with a short development distance and a weaker (more selective solvent) than employed for conventional development [8]. It is not widely used in contemporary practice.

### 6.7.3 Multiple Development

For multiple development, the thin-layer plate is developed for a selected distance, then either the layer or the mobile phase is withdrawn from the developing chamber, and adsorbed solvent evaporated from the layer before repeating the development process [8,61,67,68,71,127]. It affords a versatile strategy for separating complex mixtures. The primary experimental variables of plate length, time for development (if continuous development is used), and mobile phase composition, can be changed at any development step, and the number of steps varied to obtain the desired separation. Quantitative measurements by scanning densitometry can be made at several steps in the sequence, if desired. Consequently, it is unnecessary for all components to be separated at one time, provided they can be identified (chromatographically or spectroscopically) at different segments of the development sequence. Multiple development provides higher resolution of complex mixtures than conventional or continuous development, can easily handle samples of a wide polarity range (stepwise gradient development), and because the separated zones are usually more compact, leads to lower detection limits. The higher zone capacity than is achieved by conventional development is the result of the zone focusing mechanism (see section 6.3.7) and the optimum use of solvent

Table 6.9
Multiple development techniques

| Method | Features |
|---|---|
| Multiple chromatography | • Fixed development length<br>• Same mobile phase for each development<br>• The number of developments can be varied |
| Incremental multiple development | • Variable development length<br>(a) First development is the shortest<br>(b) Each subsequent development is increased by a fixed distance<br>(c) Last development length corresponds to the maximum useful development distance<br>• Same mobile phase for each development<br>• The number of developments can be varied |
| Increasing solvent strength gradients | • Fractionates sample into manageable subsets<br>• Optimizes separation of each subset<br>• Complete separation of all components is not achieved at any segment in the development sequence |
| Decreasing solvent strength gradients | • Uses incremental multiple development<br>• First development employs the strongest solvent with a weaker solvent for each subsequent step<br>• Final separation recorded as a single chromatogram |

selectivity. Equipment for automated multiple development is commercially available (section 6.7.5).

The main variants of multiple development are multiple chromatography and incremental multiple development, Table 6.9 [1]. Multiple chromatography is used for isocratic separations only while incremental multiple development is used for both isocratic and gradient separations. In multiple chromatography the plate is developed repeatedly over the same distance with the same mobile phase, the layer being separated from the mobile phase and dried between developments. Multiple chromatography can also be performed with a variable solvent entry position, by simultaneously incrementing the solvent entry position and solvent front position to maintain a fixed distance for each development [71,128,129]. The solvent entry position can be repositioned for each development by raising the solvent level for each development, by cutting off a portion of the bottom edge of the layer after each development, or by using a horizontal developing chamber with a layer repositioning device.

For incremental multiple development, the first development is the shortest and each subsequent development is incremented by (usually) a fixed distance or variable distances arrived at by trial-and-error, to obtain the desired separation. The last development is the longest and, in most cases, corresponds to the maximum useful development length for the layer.

Optimum conditions for isocratic multiple development are summarized in Table 6.10. The outcome of separations by multiple chromatography is the most predictable (see section 6.3.7) but is rarely the best approach for distributing sample zones throughout the whole chromatogram, and incremental multiple development is generally preferred. Incrementing the solvent entry position while simultaneously increasing

Table 6.10
Optimum separation conditions for isocratic multiple development

| Mode | Features |
|---|---|
| Multiple chromatography | • Resolution is controlled primarily by the zone center separation since zone widths are relatively constant after the first few developments<br>• The maximum zone center separation for two solutes of similar migration properties occurs when the zones have migrated 0.632 of the solvent front migration distance<br>• The maximum observed resolution is largely independent of the average $R_F$ value for the zones, although if the average $R_F$ value is small a large number of developments will be needed to reach the maximum value<br>• Compounds that are difficult to separate should be repeatedly developed with solvents that produce low $R_F$ values corresponding to the most selective mobile phase for the separation |
| Incremental multiple development | • Increment the solvent entry position to maintain a fixed distance below the least retained zone:<br>(a) Zone separation is increased owing to the longer migration distances achieved<br>(b) Zone widths are reasonably constant throughout the chromatogram<br>• Increment the solvent front migration distance:<br>(a) Avoids moving portions of the chromatogram into the solvent front region<br>(b) Resolution can be optimized throughout the chromatogram. The resolution reaches a plateau value and is not further degraded by increasing the number of developments<br>(c) The number of developments can be selected based on the requirements of the most difficult zones to separate |

the development distance in successive developments provides increased resolution of sample zones largely because of the increase in the zone migration distance. The normal zone broadening mechanism, leading to wider zones, is effectively counteracted by the focusing mechanism. Interestingly, resolution in incremental multiple development reaches a plateau value and is not subsequently reduced by increasing the number of development steps, unlike multiple chromatography.

Isocratic incremental multiple development with changes in the solvent entry position is selected for the separation of compounds with similar retention that tend to migrate as a compact group (e.g. isomers, diastereomers and analogs with minor structural variations). Separation is usually obtained only in the higher regions of the layer, and at each subsequent development, a significant amount of time is wasted while the advancing solvent front reaches the level of the lowest zone. As well as increasing the separation time, this process results in difficult separations being attempted in regions of the layer where the mobile phase velocity is inadequate to minimize zone broadening. An example of the separating power of different isocratic multiple development approaches is illustrated in Figure 6.9 [68]. The poorest separation is obtained by multiple chromatography. The separation is improved by using incremental multiple development, with the best separation (nearly to baseline) achieved by the simultaneous change of the solvent entry and solvent front migration distance at each successive development. This minimizes zone broadening and enhances the zone center separations by migrating the sample components over a longer plate length while

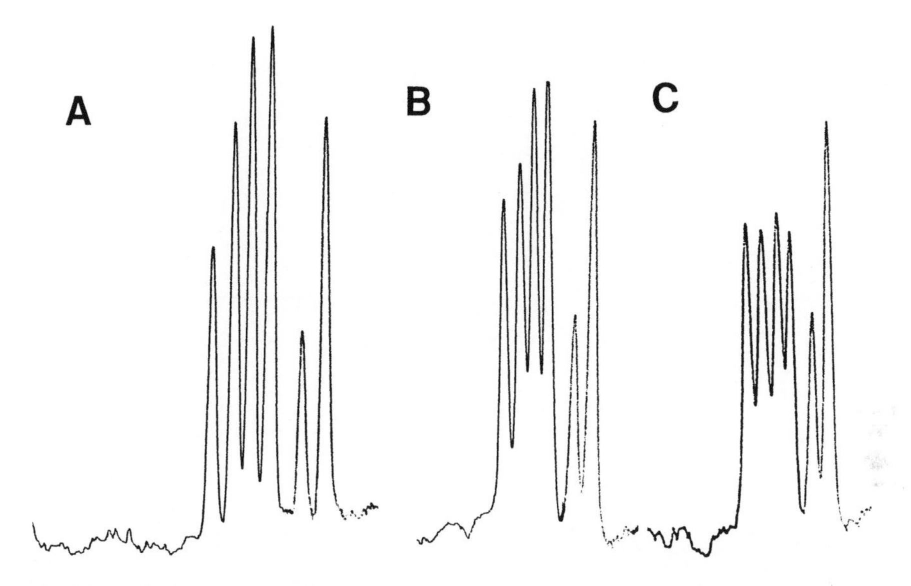


Figure 6.9. Separation of a mixture of estrogens by incremental multiple development with variable (A) and fixed (B) solvent entry position, and by multiple chromatography (C). A nine step sequence with the mobile phase cyclohexane-ethyl acetate (3:1, v/v) was used for (A) and (B) and seven 7 cm developments for (C), all on a silica gel HPTLC layer. The estrogens, in order of migration, are 17β-dihydroequilenin, 17α-dihydroequilenin, 17β-estradiol, 17α-estradiol, equilenin and estrone. (From ref. [68]. ©Research Institute for Medicinal Plants).

maintaining a mobile phase velocity that closely approaches the most favorable value for the separation.

For samples containing components of a wide retention range, some form of gradient development is required to obtain a separation of all components either in a single chromatogram or in separate chromatograms for successive developments. Figure 6.10 illustrates one such example [68,130]. The mobile phase can be optimized to separate either the low oligomer number or high oligomer number components in a single development, but not both simultaneously. Using multiple development with a stepwise mobile phase gradient, all of the oligomers are separated in a single chromatogram. Continuous composition gradients are rarely used in thin-layer chromatography, unlike column liquid chromatography. They require special equipment and less convenient experimental conditions than step gradients. In addition, step gradients can be constructed to mimic a continuous gradient with the added advantage that the zone focusing mechanism can be exploited to minimize zone broadening. For forced flow, which closely resembles column liquid chromatography, continuous composition gradients are commonly used and multiple development rarely.

Common methods for constructing solvent strength gradients in multiple development are summarized in Table 6.9. Gradients of increasing solvent strength are used to fractionate complex mixtures by separating just a few components in each

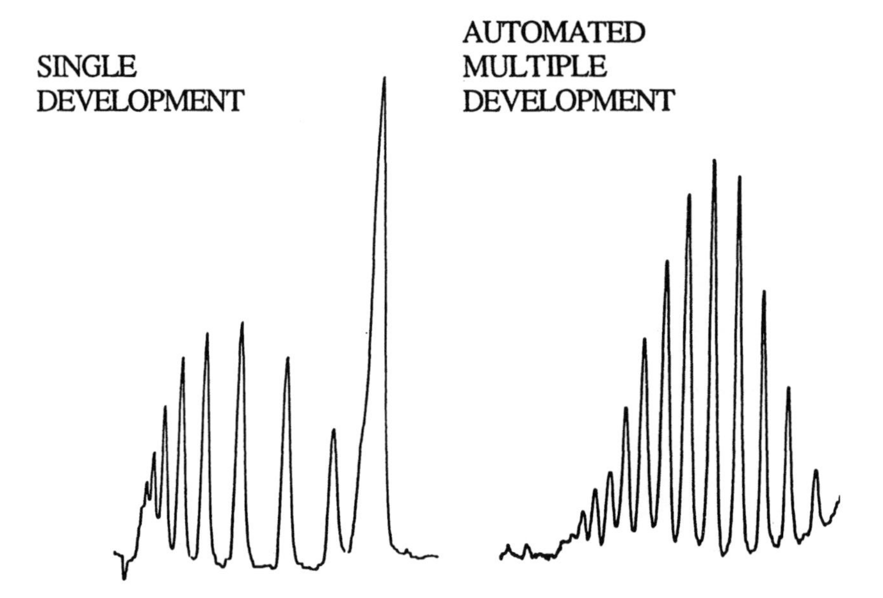


Figure 6.10. Comparison of normal development and incremental multiple development with a decreasing solvent strength gradient (AMD) for the separation of poly(ethylene glycol) 400 as its 3,5-dinitrobenzoate ester. The AMD separation employed a 15- step gradient with methanol, acetonitrile and dichloromethane as solvents. (From ref. [68]. ©Research Institute for Medicinal Plants).

step [8,71,131]. Individual compounds are usually identified and quantified by scanning intermediate steps at which the sample components of interest are separated. In this way the zone capacity can be much greater than predicted for a complete separation recorded as a single chromatogram. On the other hand, this approach can be tedious when many components are of interest, and it is difficult to automate. The alternative strategy is based on incremental multiple development with separation of the sample for the shortest distance in the strongest mobile phase with each subsequent, longer development using mobile phases of decreasing solvent strength. This strategy is most useful when it is desired to record the separation as a single chromatogram. On the other hand, the peak capacity is limited because all components must fit between the position of the sample origin and the final solvent front. The two approaches for exploiting solvent strength gradients are complementary, and a selection is made based on the properties of the sample. The decreasing solvent strength gradient approach is the operating basis of the automated multiple development chamber (section 6.7.5).

### 6.7.4 Two-Dimensional Development

For two-dimensional thin-layer chromatography the sample is spotted at the corner of the layer and developed along one edge of the plate [61,71,127,132,133]. The solvent is then evaporated, the plate rotated through 90°, and redeveloped in the orthogonal

Table 6.11
Methods for generating two different retention mechanisms in orthogonal directions

| |
|---|
| • Develop in orthogonal directions with two solvent systems exhibiting different selectivity for the sample components. |
| • Use a bilayer plate prepared from two sorbents with different selectivity for the sample. The sorbent layer for the first development is a narrow strip that abuts the much larger area used for the second development. Commercially available plates have silica gel and revered-phase layers as adjacent zones. |
| • Use a layer prepared from a mixture of two different sorbents and develop with different mobile phases such that the retention mechanism for the two, orthogonal, developments is governed by the properties of one of the sorbents in each direction. |
| • After the first development the selectivity of the layer is changed by impregnation with a chemical reagent or immiscible solvent prior to the second development. |
| • After the first development the properties of the sample are modified by chemical reaction or derivatization prior to the second development. |

direction. If the same solvent is used for both developments then the sample will be redistributed along a line from the corner at which the plate was spotted to the corner diagonally opposite. In this case, only a small increase in resolution can be anticipated corresponding to the greater migration distance for the sample. The realization of a more efficient separation system implies that the resolved sample is distributed over the entire plate surface. This can be achieved only if the selectivity of the separation mechanism is complementary in the orthogonal directions. The possibility of achieving a large zone capacity for these conditions, Table 6.4, sustains interest in this technique [1,134,135].

Some potential methods for generating two different retention mechanisms in orthogonal directions are summarized in Table 6.11 [61,132,136-139]. Using two solvent systems with complementary selectivity is the simplest approach but is often only partially successful. In many cases the two solvent systems differ only in their intensity for a given set of intermolcecular interactions, and are not truly complementary. Chemically bonded layers can be used in the reversed-phase and normal-phase mode and allow the use of additives and buffers as a further means of adjusting selectivity. Aminopropylsiloxane-bonded layers can be used as an ion-exchange system with acidic mobile phases and as a reversed-phase or normal-phase system with a neutral or basic mobile phase. In several reports as many as 20 to 30 components were successfully separated by two-dimensional thin-layer chromatography indicating the potential of this approach for analyzing complex mixtures, even, as seems likely, the conditions used were not truly optimum. Also, the potential increase in zone capacity obtainable by using multiple development or forced flow to minimize zone broadening and the use of solvent strength gradients, as part of the optimization strategy, seem to have largely gone unrecognized [1,61].

The acceptance of two-dimensional thin-layer chromatography for quantitative analysis rests on providing a convenient method for *in situ* detection, quantification, and data analysis. Slit-scanning densitometers are designed for lane scanning and are not easily adapted for area scanning. Adaptation to scanning two-dimensional separations was demonstrated using normal scanning operations with small steps between scans or by zigzag scanning. Special software is required to map the layer surface and define zone locations and their optical density as three-dimensional plots or contour diagrams [61,140,141]. The awaited breakthrough in general detection for two-dimensional planar separations is likely to come from optical imaging systems (see section 6.9.4).

### 6.7.5 Developing Chambers

The development process in thin-layer chromatography can be carried out in a variety of chambers that differ significantly in design and sophistication [9,32,36,124]. For convenience these are often categorized as normal (N-chamber) and sandwich (S-chamber) type chambers, and further subdivided based on whether the internal atmosphere is saturated ($N_S$ or $S_S$) or unsaturated ($N_U$ or $S_U$). Normal chambers have a depth of gas phase in front of the layer greater than about 3 mm. For sandwich layers the gas phase is less than 3 mm. Saturation of the vapor phase is achieved by using solvent saturated pads as a chamber lining or, for sandwich cambers, a second solvent filled layer facing the separation layer. The mobile phase velocity and $R_F$ values can be different for different chamber designs using the same stationary and mobile phases. To compare results from different chamber types, it is necessary to determine the chamber saturation grade [36].

The twin-trough chamber is the simplest of the developing chambers for conventional and high performance thin-layer chromatography [142]. It consists of a standard rectangular tank with a raised, wedge-shaped bottom. The wedged bottom divides the tank into two compartments, so that it is possible to either develop two plates simultaneously, or to use one compartment to condition the layer prior to development. The plate is placed in one compartment and mobile phase or conditioning solvent in the other. The mobile phase can be added directly to the plate compartment or, when the mobile phase is used for conditioning as well, the chamber can be tilted to allow transfer of mobile phase into the compartment for development. The twin-trough chamber is widely used for routine quality control applications.

The horizontal development chamber can be used in either the normal or sandwich configuration for either conventional edge-to-edge or simultaneous edge-to-center development [124,129,143]. Starting the development simultaneously from opposite edges enables the number of sample separated to be doubled with the same separation time. The mobile phase is transported from the reservoirs to the layer by two glass slides; the liquid rises by surface tension and capillary forces. The mobile phase then travels through the layer by capillary action. When a plate is developed from both edges simultaneously, the chamber must be leveled to allow the two solvent fronts to migrate at the same speed and meet precisely in the middle. The sandwich configuration of the horizontal developing chamber is unsuitable for mobile phases containing volatile acids or bases, or mixtures with a large concentration of a volatile polar solvent, such as

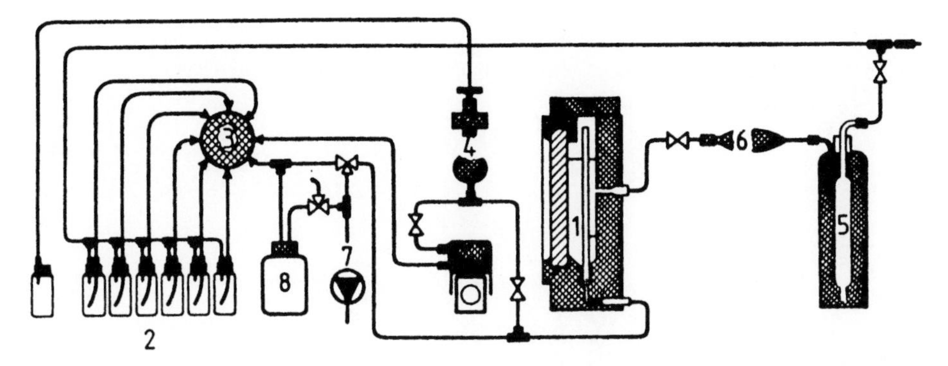


Figure 6.11. Schematic diagram of the automated multiple development chamber. Identification: 1 = developing chamber; 2 = solvent reservoirs; 3 = solvent selection valve; 4 = solvent mixer; 5 = wash bottle for preparation of the gas phase for layer conditioning; 6 = gas phase reservoir; 7 = vacuum pump; and 8 = solvent waste reservoir.

methanol or acetonitrile. The Vario-KS chamber is similar to the horizontal developing chamber [32] but allows the layer to be segregated into six individual lanes, each with its own conditioning trough and mobile phase reservoir. It allows up to six different separation conditions to be evaluated simultaneously, and is used primarily as a scouting tool to optimize separation conditions.

Automation of conventional development is possible using an automated developing chamber [9]. Preconditioning, normal or sandwich configuration, mobile phase migration distance, and the drying conditions are selectable from a microprocessor-based control unit. Mobile phase is automatically added to the developing chamber and development occurs under controlled conditions in the usual way. A sensor monitors the solvent-front migration distance. When the selected solvent-front migration distance is reached, solvent is drained from the chamber, and the layer is dried with filtered cold or warm air.

A schematic diagram of the automated multiple development chamber is shown in Figure 6.11 [61,67,124,144]. This chamber was developed for automated separations requiring incremental multiple development with a decreasing solvent strength gradient. The operating variables of layer conditioning, solvent-front migration distance, mobile phase composition for each development step, drying time, and the number of developments are entered into the computer-based control unit. Without further intervention the complete separation sequence is carried out. The layer remains in a fixed position and the mobile phase components are selected, mixed, and then delivered to the chamber. A sensor monitors the solvent-front migration distance. The development is stopped when the solvent front reaches the selected distance, mobile phase drained from the chamber, and mobile phase trapped by the layer removed by vacuum. A new development sequence is then initiated until the complete program is executed. Typically each development is 3-5 mm longer than the previous one. The first few developments with a strong solvent are used primarily to focus the sample zones.

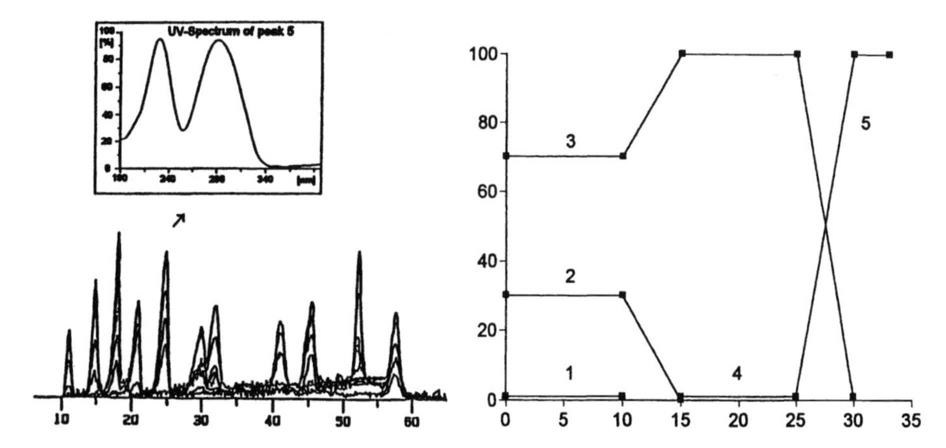


Figure 6.12. Separation of a mixture of polar crop-protecting agents on silica gel by automated multiple development. Shown are the gradient profile used for the separation, the use of multiple wavelength scanning for detection and an *in situ* UV spectra for one peak (≈ 50 ng). Solvent compositions for the gradient: 1 = aqueous ammonia; 2 = acetonitrile; 3 = dichloromethane; 4 = formic acid; and 5 = hexane. (Adapted from ref. [21]; ©American Chemical Society).

Subsequent steps are longer with weaker solvents and responsible for the separation. The number of development steps depends on the complexity of the mixture and the shape of the desired gradient profile. From 10-30 developments requiring 1.5 to 4.5 h for completion are not unusual. A typical separation of crop protecting agents employing a gradient containing both acid and base modified organic solvents, is shown in Figure 6.12 [20,21]. Screening for 265 pesticides in water with typical detection limits in the 10s of nanogram range per component (or about 50 ng / l for a 500 ml water sample prepared by solid-phase extraction) is possible. Multiple-wavelength scanning is used to improve the certainty of identification based on simultaneously matching of migration distances and absorbance ratios to those of known substances. Full *in situ* UV spectra can also be searched against appropriate libraries or matched with standards run on the same plate.

Some samples are unsuitable for separation by multiple development [67,68,71]. Compounds with significant vapor pressure may be lost during the repeated solvent evaporation steps performed under vacuum. Artifact peaks from unstable compounds may be mistaken for sample components. Compounds that are easily oxidized, photolyzed, or hydrolyzed to more polar products yield two separated zones in each development. In a multi-step program, the accumulative effect can result in a complex chromatogram from a single component. Solvents of low volatility and/or high polarity should be avoided, since they are only slowly removed from the layer by vacuum and considerably extend the total separation time. Solvent impurities can be a source of ghost peaks (identifiable by scanning between sample lanes).

Two general approaches are used for forced flow separations. Rotational planar chromatography uses centrifugal forces created by spinning the plate around a central axis to move the mobile phase through the layer [31,42]. The rotation speed can be varied from 80 to 1,500 rpm. By scrapping out zones in the layer, either circular, anticircular or linear development is possible. A collection system fixed to the rotor allows separated fractions to be collected as they leave the layer when operated in the circular mode for preparative-scale applications. The overpressured development chamber eliminates the vapor space above the layer by contacting the layer surface with a plastic membrane pressurized on its opposite surface by hydraulic pressure of up to 50 atmospheres [12,45,46]. A mechanical pump forces the mobile phase through the layer at the selected velocity. Special layer preparation is required for linear development. The sorbent is removed from the bottom (sample application end) and two sides of the layer to a distance of about 5 mm and the edges sealed with latex glue [48]. This ensures that the mobile phase flow occurs only in the desired direction. An eluent trough scored into a polymeric sheet inserted between the hydraulic cushion and the layer is used to create a linear solvent front. A second insert at the exit position is required if the chamber is used in the elution mode with on-line detection, similar to normal operation in column liquid chromatography. Since only a single sample can be separated with on-line detection, this is normally used to identify fractions for collection in preparative-scale chromatography or coupling to other separation or identification techniques.

## 6.8 METHOD DEVELOPMENT

Since the zone capacity in thin-layer chromatography is limited (Table 6.4), the number of detectable components in the mixture and their range of polarity is used to determine the development technique. For capillary flow separations, in a single development, mixtures containing more than 8 to 10 components may be too difficult or impossible to separate. In addition, if the range of polarities is too wide then multiple development techniques using a mobile phase gradient will be necessary. Of course, all components of a mixture may not be of equal interest and then the goal of the method is to separate the components of major interest from each other and from the less important components as a group. Method development will be aided if standards for the relevant compounds are available. Standards simplify zone tracking and enable detection characteristics and the possibility of spectroscopic resolution of incompletely separated zones to be established. Standards are also required for calibration, if quantification is required, and to construct spectral libraries for identification purposes. The expected concentration range of relevant compounds may indicate the need for derivatization to obtain required detection limits.

A general guide to stationary phase selection for separations by thin-layer chromatography is summarized in Figure 6.13 (see section 6.4). Silica gel is the most widely used stationary phase for thin-layer chromatography and is generally the first choice for the separation of low molecular mass organic compounds soluble in common organic solvents. Most thin-layer sorbents are small pore materials optimized for the separation of compounds with molecular weights below about 750. Precipitation chromatography is

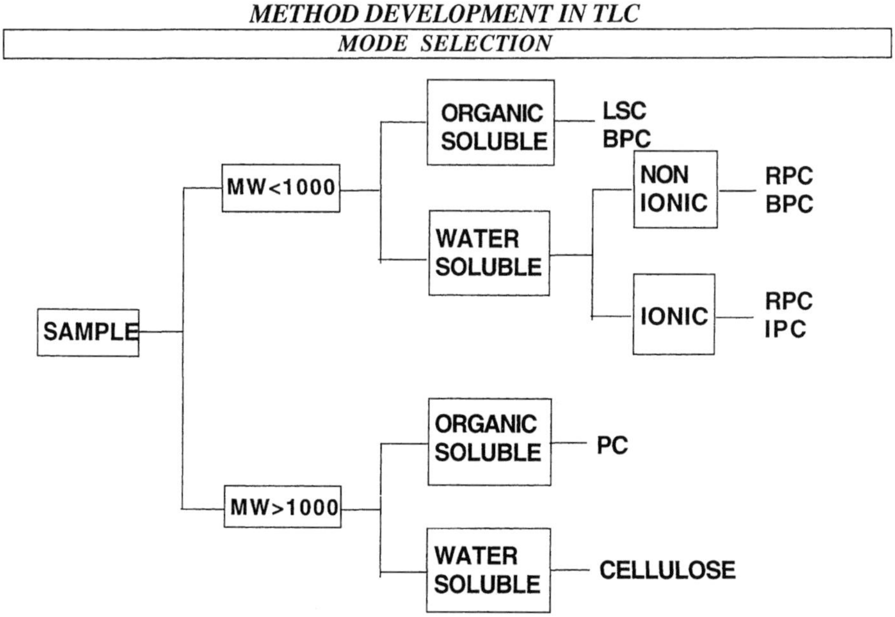

METHOD DEVELOPMENT IN TLC

MODE SELECTION

SAMPLE
MW<1000
ORGANIC SOLUBLE — LSC BPC
WATER SOLUBLE
NON IONIC — RPC BPC
IONIC — RPC IPC
MW>1000
ORGANIC SOLUBLE — PC
WATER SOLUBLE — CELLULOSE

Figure 6.13. Separation mode selection guide for TLC. LSC = liquid-solid chromatography on an inorganic oxide adsorbent; BPC = liquid-solid chromatography on a chemically-bonded sorbent; RPC = reversed-phase chromatography with a water-containing mobile phase and chemically-bonded stationary phase; IPC = ion-pair chromatography with reversed-phase separation conditions; and PC = precipitation chromatography. (From ref. [151]; ©Elsevier)

a unique application of thin-layer chromatography to the separation of high molecular mass polymers based on solubility differences in a mobile-phase composition gradient [145,146]. For the separation of water soluble, high-molecular-mass biopolymers cellulose layers are about the only useful choice. Reversed-phase chromatography on chemically bonded layers is generally used for the separation of compounds difficult to separate on silica gel, Figure 6.14. Compounds of low polarity can be difficult to separate on silica gel because of weak retention (mobile phase selection is limited because most solvents are too strong for these separations) and polar compounds because of strong retention (mobile phase selection is limited because most solvents are too weak for these separations). Ions and easily ionized compounds are frequently separated by reversed-phase chromatography using buffered mobile phases (weak acids and bases) or ion-pair reagents (strong acids and bases). There are only a limited number of stationary phases suitable for ion-exchange chromatography. Except for 3-aminopropylsiloxane-bonded layers, which function as a weak anion exchanger with an acidic mobile phase, ion exchange is not a widely used separation mechanism in thin-layer chromatography.

Given the similarity of retention mechanisms, it is hardly surprising that the principal methods of mobile phase selection in thin-layer chromatography are similar to those

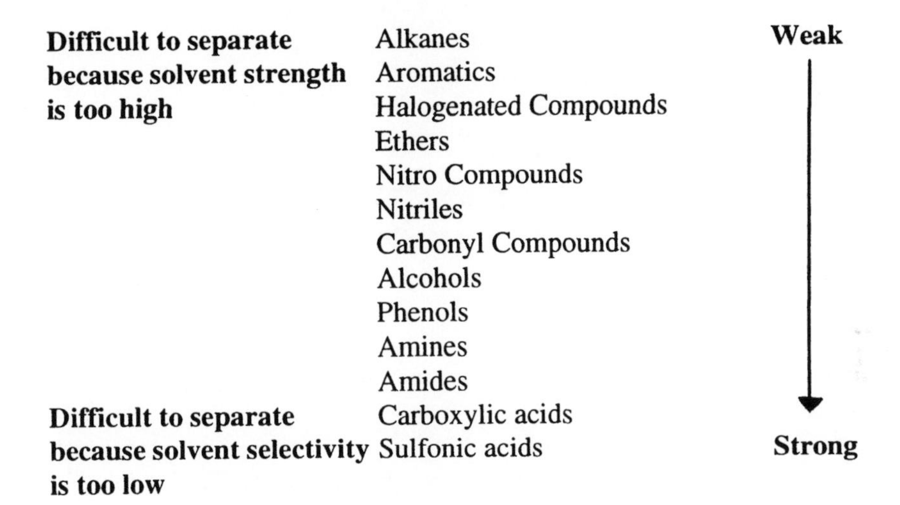


Figure 6.14. General adsorption scale for separations by silica gel thin-layer chromatography.

followed in column liquid chromatography (section 4.4.1). Since the mobile phase is evaporated prior to detection, a wider range of UV absorbing solvents are commonly used in thin-layer chromatography than is the case for column liquid chromatography. Solvents must be of high purity, since involatile impurities and stabilizers intentionally added to some solvents, remain adsorbed to the layer causing sloping and unstable baselines in scanning densitometry [86]. The most significant difference between column and thin-layer methods is that in thin-layer chromatogaphy equilibrium is not necessarily established throughout the separation. When multi-component mobile phases are used, composition gradients can form in the development direction due to demixing (see section 6.3.2). Demixing effects are less apparent when saturated developing chambers are used. The presence of a vapor phase in TLC further complicates matters with both migration distances and migration order influenced by the level of saturation of the development chamber [49]. These considerations hinder optimization strategies based on the composition of the solvent added to the developing chamber as popularized in column liquid chromatography.

The selection of a mobile phase for the separation of simple mixtures may not be a particularly difficult problem and can be arrived at quite quickly by guided trial and error experiments [9,147-152]. A mobile phase with the correct strength for a single development will migrate the sample into the $R_F$ range 0.2-0.8, or thereabouts. If of the correct selectivity, it will distribute the sample components evenly throughout this range, or meet some other resolution criterion established for the particular separation. Solvent systems can be screened in parallel using either several development chambers or a device like the Vario-KS chamber (section 6.7.5). Alternatively, sample spots (up to 16 on a high performance layer) are applied at suitable positions on a single layer and automatically developed in sequence with 45 μl of solvent or solvent mixtures using an automated sample applicator [150,153]. The individual circular chromatograms enable rapid identification of solvents with suitable strength and selectivity for the separation. The three-point window diagram [153] and simple mixture design approaches [150,154,155] are other methods suitable for the quick identification of separation conditions for simple mixtures. However, whenever the number of components in a mixture exceeds all but a small fraction of the zone capacity for the separation system, a more systematic method of mobile phase optimization is required.

### 6.8.1 Prisma Model

The Prisma model is widely used for optimization of mobile phase composition in thin-layer chromatography [156,157]. Application of the model proceeds in three stages: (i) selection of the chromatographic system, (ii) optimization of the selected mobile phase components, and (iii) selection of the development method. The optimization procedure starts with silica gel, since this is the most commonly used stationary phase for thin-layer chromatography, although the method is equally applicable to normal-phase and reversed-phase separations on chemically bonded layers. Suitable solvents for the separation are identified in an initial experimental screening step. These experiments are carried out on thin-layer plates in unsaturated chambers with single solvents chosen from the different selectivity groups according to Snyder (Table 4.14, section 4.4.1). After these initial experiments, the solvent strength is adjusted so that the sample components are distributed in the $R_F$ range 0.2-0.8. If the substances migrate into the upper third of the plate the solvent strength is reduced by dilution with hexane, the strength adjusting solvent (solvent strength = 0). If the substances remain in the lower third of the plate with the single solvents their solvent strength has to be increased by addition of a strong solvent such as water or acetic acid. A similar procedure is followed for reversed-phase chromatography using water miscible solvents and water as the strength adjusting solvent (solvent strength = 0). From these trial experiments, mobile phases giving the best separation (maximum number of separated zones) are selected for further optimization in the second part of the model. If it is anticipated that forced flow will be used for the separation then it is advantageous to include one solvent in the selected solvent systems in which the sample does not migrate. This solvent can be used in a prerun to eliminate the disturbing zone caused by undissolved gases in the sealed layer (section 6.3.3).

Between two and five solvents can be selected for mobile phase optimization. Modifiers such as acids, ion-pair reagents, etc., can be added to improve the separation and reduce tailing. Modifiers are generally used in a low and constant concentration so that their influence on solvent strength can be neglected. The Prisma model, Figure 6.15, is a three dimensional geometrical design which correlates the solvent strength with

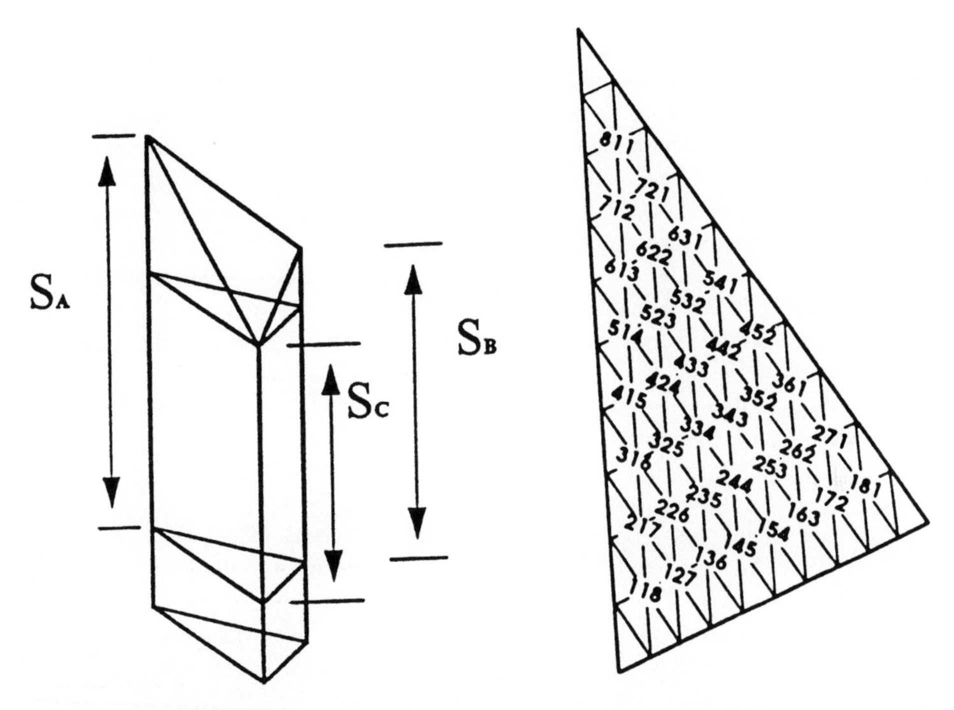


Figure 6.15. The Prisma mobile phase optimization model showing the construction of the prism and the selection of selectivity points.

the selectivity of the mobile phase [156]. The model consists of three parts: the base or platform representing the modifier; the regular part of the prism with congruent base and top surfaces; and the irregular truncated top prism (frustum). The lengths of the edges of the prism ($S_A$, $S_B$, $S_C$) correspond to the solvent strengths of the single solvents (A, B and C). Since the selected solvents usually have different solvent strengths, the edges of the prism may be of unequal length and the top plane of the prism will not be parallel and congruous with its base. Cutting the prism parallel to its base at the height of the lowest edge (determined by the solvent strength of the weakest solvent, solvent C in Figure 6.15), gives a regular prism, where the top and any planes representing weaker solvents diluted with a strength adjusting solvent are parallel equilateral triangles. The upper frustum is used for mobile phase optimization of polar compounds in normal-phase chromatography, while the regular part is used for the separation of non-polar and moderately polar substances. For reversed-phase chromatography, the regular part of the prism is used to optimize the separation of both non-polar and polar substances.

For polar compounds optimization is always started on the top irregular triangle of the model, either within the triangle, when three solvents are selected, or along one side, for binary mobile phases. Any solvent composition on the face of the triangle can be represented by a three-coordinate selectivity point ($P_S$); each coordinate corresponding to the volume fraction of the solvent at that position on the triangle, Figure 6.15. Optimization is commenced by selecting solvent combinations corresponding to the center point $P_S$ = 333 and three other points close to the apexes of the triangle $P_S$ = 811, 181 and 118. If the separation obtained is insufficient other selectivity points are tested around the solvent combination that gave the best separation. On changing the selectivity points on the top triangle the solvent strength changes as well, especially when the solvent strengths of the solvents used to construct the prism are considerably different. The strength of the solvent should be adjusted with the strength adjusting solvent to maintain the separation in the optimum $R_F$ range. It may also be advisable to change the selectivity points by small increments if regular step sizes cause large changes in resolution. To aid optimization experimental data can be fitted to a 3-dimensional retardation surface with x- and y-coordinates as selectivity points and the z-coordinate as the $R_F$ value [158,159].

The regular center portion of the prism is used to optimize the mobile phase composition for the separation of non-polar and moderately polar compounds. The initial solvent composition corresponds to the center of the triangular top face of the regular prism ($P_S$ = 333); this composition is then diluted to bring all sample components into the $R_F$ range 0.2-0.8. At this solvent strength three more chromatograms are run corresponding to the selectivity points close to the apexes of the triangle. These initial runs are then used to choose selectivity points for further chromatograms until the best solvent composition is located. For saturated developing chambers there is a linear relationship between $R_F$ values and the solvent strength at a constant $P_S$ value ($\ln R_F = d(S_T) + e$, where $S_T$ is the solvent strength and d and e are regression constants). At a constant solvent strength there is a quadratic relationship between the $R_F$ value and the selectivity points describing the retardation surface [$R_F = a\,(P_S)^2 + b\,(P_S) + c$, where a, b and c are regression constants). These relationships can form the basis of a computer-aided optimization strategy according to a fixed experimental design requiring 18 experiments [152] or a general approach with decisions based on the interpretation of the retardation surfaces [159,160]. Within the concept of the Prisma model, selection of the correct solvent strength is referred to as vertical optimization, and identification of the optimum mobile phase composition at a fixed solvent strength as horizontal optimization.

Optimization of the solvent strength by varying the selectivity points is carried out until the required separation is obtained. If no adequate separation is obtained then a different layer or additional solvents must be selected and the new system optimized by the previous procedure. Nearly adequate separations can be improved in the third part of the Prisma model by selecting a different development mode. If an increase in efficiency is required to improve the overall separation then forced flow methods should be used. If the separation problem exists in the upper $R_F$ range then anticircular development may be the best choice, if in the lower $R_F$ range, then circular development is favored.

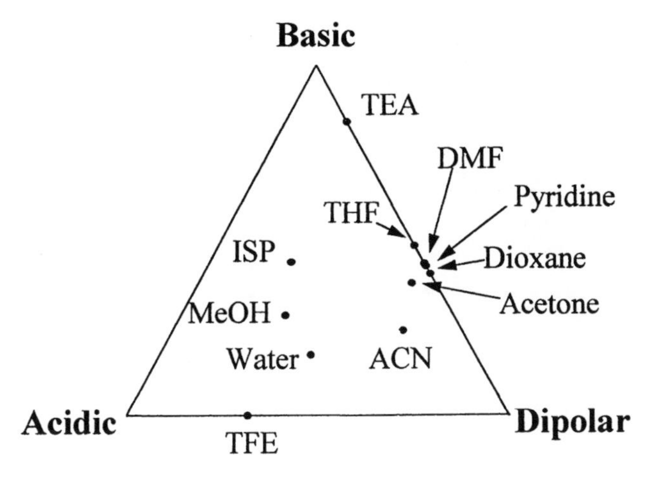


Figure 6.16. Solvent-selectivity triangle based on normalized solvatochromic parameters for some common water miscible organic solvents. TEA = triethylamine, THF = tetrahydrofuran, DMF = N,N-dimethylformamide; ISP = 2-propanol; MeOH = methanol; ACN = acetonitrile; and TFE = 2,2,2-trifluoroethanol. (From ref. [95]; ©Research Institute for Medicinal Plants).

### 6.8.2 Solvation Parameter Model

The solvation parameter model has been used as the basis of a structure-driven retention model for method development in reversed-phase thin-layer chromatography [95,151,161-164]. The model should be applicable to normal-phase separations on chemically bonded layers as well, but not silica gel layers [164]. Solute size differences and site-specific interactions on silica gel are not adequately accounted for by the model, which results in poor predictions of retention. The solvation parameter model is described in section 1.4.3. The $R_M$ value (section 6.3.1) is used as the dependent variable.

The selection of mobile phase components in reversed-phase chromatography is restricted to solvents that are miscible with water. These solvents can be classified according to their capability for different intermolecular interactions based on their solvatochromic parameters (section 4.4.1). A visual indication of solvent selectivity differences is obtained by plotting the solvents on the surface of a triangle using normalized coordinates, Figure 6.16 [95]. The most selective solvents are found along the edges and towards the apexes for the indicated interactions with those solvents containing a blend of properties located on the face. The selectivity range that can be achieved with the three common solvents methanol, acetonitrile, and tetrahydrofuran is rather limited as indicated by their position on the triangle compared to the greater selectivity space available if all solvents are considered. It is also clear from the distribution of solvents on the triangle that solvents, which are strong hydrogen-bond bases with weak dipole-type properties, are uncommon. Triethylamine is the best of these solvents but is immiscible with water and could only be used as a component of a ternary mobile phase containing a polar organic solvent. Acetone, acetonitrile, 2-

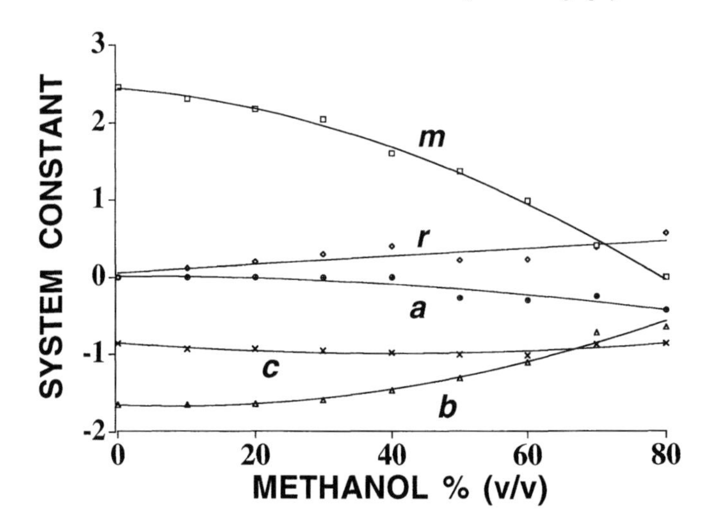


Figure 6.17. A system map for a cyanopropylsiloxane-bonded layer with methanol-water as mobile phase. The *m* system constant is a measure of the difference in cohesion and dispersion interactions between the solvated layer and the mobile phase; the *r* system constant the difference in capability for electron lone pair interactions; the *a* system constant the difference in hydrogen-bond basicity; the *b* system constant the difference in hydrogen-bond acidity; *c* is the model constant and among other contributions contains the phase ratio; and the *s* system constant representing differences in dipole-type interactions is zero at all mobile phase compositions.

propanol, methanol and 2,2,2-trifluoroethanol provide a convenient range of hydrogen-bond acidity. Methanol and 2-propanol are simultaneously strong hydrogen-bond bases and dipolar, and for this reason are found towards the center of the triangle. They could be considered as polar solvents of low selectivity. 2,2,2-Trifluoroethanol is the strongest hydrogen-bond acid (stronger than water) and has zero hydrogen-bond basicity. It is quite dipolar but considerably more selective than the other alcohols for hydrogen-bond interactions. Acetonitrile (or dioxane), acetone (or tetrahydrofuran), and pyridine (or N,N-dimethylformamide) provide a reasonable range of hydrogen-bond base properties. Acetonitrile is the most dipolar solvent with minimal (although significant) capacity for hydrogen-bond interactions. The six water miscible organic solvents (acetone, acetonitrile, 2-propanol, methanol, 2,2,2-trifluoroethanol and pyridine) are expected to provide a reasonable range of separation selectivity for reversed-phase separations.

Method development for binary mobile phases using the solvation parameter model is based on the use of system maps. A system map is a continuous plot of the system constants obtained from experimental data fit to the solvation parameter model against mobile phase composition. However, once constructed the system map is a permanent record of system properties. It is used in all calculations and is not restricted to the compounds used to construct the system map. A typical system map for methanol-water mobile phases on a 3-cyanopropylsiloxane-bonded layer is shown in Figure 6.17. System maps for several binary mobile phases on octadecylsiloxane-bonded [95],

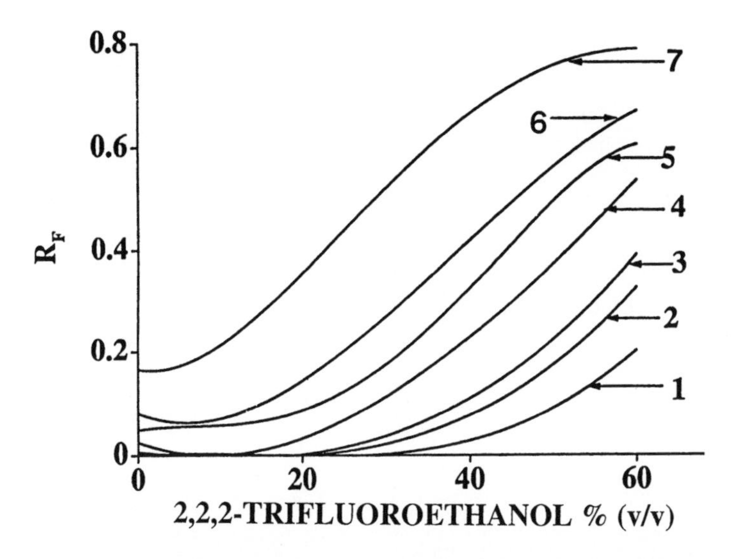


Figure 6.18. Retention map for the separation of analgesics by reversed phase thin-layer chromatography on an octadecylsiloxane-bonded layer with 2,2,2-trifluoroethanol-water mixtures as mobile phase. Compounds: 1 = chloropheniramine; 2 = ibuprofen; 3 = naproxen; 4 = phenacetin; 5 = aspirin; 6 = caffeine; and 7 = acetaminophen.

cyanopropylsiloxane-bonded [162], and spacer bonded propanediol [151] layers are available.

Retention maps are created from the system maps for all compounds to be separated. This is conveniently done using a spreadsheet for the calculations and graphics for evaluation. All mobile phase and stationary phase combinations for which system maps are available can be compared in the search for the optimum system. A typical retention map for the separation of a mixture of analgesics on an octadecylsiloxane-bonded layer with 2,2,2-trifluoroethanol-water as a mobile phase is shown in Figure 6.18 [95]. Those solvent compositions resulting in acceptable zone separation are easily identified by visual inspection. Computer simulation of retention maps allows those systems likely to provide an acceptable separation to be identified before experimental work commences. Consequently, this approach is able to direct experimental procedures and not just simply aid in their interpretation. The agreement between model predicted and experimental $R_F$ values is generally good. Differences are typically less than 0.05 $R_F$ units. Retention surfaces for the optimization of ternary solvent systems based on a mixture-design approach combined with the solvation parameter model have also been demonstrated [164].

### 6.8.3 Computer Simulations

The solvation parameter model (section 6.8.2) can be considered a suitable model for computer-aided method development. Computer-aided strategies for mobile phase optimization using window diagrams, overlapping resolution maps, simplex methods, and iterative procedures have found occasional use in thin-layer chromatography, but not to the same extent as for column chromatography [9,147-149,151,165]. In these procedures, some form of statistical design is used to select a group of solvents for evaluation, or alternatively, the results obtained from an arbitrary selection of solvents are compared to indicate the best separation. In the simplex method an evolving experimental design is used to predict new mobile phase compositions from initial solvent compositions guided by a set of rules that (hopefully) direct the movement of the simplex to a solvent composition providing an acceptable separation. To rank separation quality for computer interpretation a single-value numerical index, or a technique, such as overlapping resolution maps for visual evaluation, is used [9,166]. Many mathematical functions are used to characterize the quality of thin-layer separations but none can be considered ideal. The choice of the separation quality index can lead to the prediction of different optimum mobile phase compositions for a separation, which is disconcerting. Resolution based functions favored in column chromatography (section 1.6.2) are not useful in thin-layer chromatography because zone widths are difficult to predict and depend on migration distance. Typically, separation quality is defined by an index based on the zone center separation, sometimes combined with a function to characterize the zone distribution throughout the separation.

Computer simulation of separations for binary mobile phases containing a single strong solvent are usually based on one of the functions $R_M = a \log (X_S) + b$ or $R_F = a (X_S)^2 + b (X_S) + c$ where a, b and c are regression constants and $X_S$ is the mole fraction of strong solvent [167]. A certain amount of experimental data is required to determine the regression coefficients after which additional $R_F$ values can be estimated by interpolation. This approach can only estimate results for compounds included in the initial experiments. Results for simulation of two-dimensional separations based on sequential application of the above equations for one-dimensional separations varied from poor to reasonable without obvious reasons for the variation [61,168,169].

### 6.8.4 Gradient Methods

Mobile phase composition gradients are required for the separation of samples containing components of a wide retention range. Stepwise solvent strength gradients are easier to implement than continuous gradients in thin-layer chromatography, especially when multiple development is used for the separation (section 6.7.3). Soczewinski's group described a modification of the horizontal developing chamber for stepwise gradient development with increasing solvent strength mobile phases [170,171]. The modified design of the solvent delivery system allows the mobile phase to be adsorbed to the last drop. Therefore, successive volumes of mobile phase of increasing solvent strength can be introduced after complete adsorption of the previous fraction by the layer. The total of all volume fractions of each mobile phase used to construct the gradient is usually equivalent to the hold-up volume of the layer. Solvent consumption is minimal and any gradient program, including continuous or multiple component gradients can be generated. The real gradient profile in the layer, however, may be different to the programmed gradient due to solvent demixing and the slow exchange of stagnant mobile phase in the sorbent pores with the mobile phase of higher solvent strength.

Multiple gradient development (MGD) employs successive development of the layer over decreasing distances with mobile phases of increasing solvent strength [172,173]. The layer is dried between developments. Each development distances is selected such that the position of the most advanced zones separated in the first few developments remain fixed in position by locating the new solvent front at a level below them. The method is easily performed in the horizontal developing chamber with minimal solvent use. The most important approach to stepwise gradient development uses multiple development over increasing migration distances with a decreasing solvent strength gradient. This is the basis of the automated multiple development chamber (section 6.7.3).

Suitable models for the calculation of migration distances in stepwise gradient development using eluent fractions of increasing solvent strength without interruption of the development process [171,174] and by incremental multiple development with increasing and decreasing solvent strength gradients [172,173,175-177] have been described. The general approach is similar in all cases. The relationship between the $R_F$ value and mobile phase composition for each mobile phase used to construct the gradient is determined experimentally and fit to a mathematical function as described in section 6.8.3. For computer simulation, the number of developments and the composition of the mobile phase and solvent-front migration distance for each development are selected. The computer then calculates the migration distance for each component by considering each development step separately and updating the zone origin for each component based on its location in the previous development. Computer simulation enables the influence of solvent composition, number of developments and solvent-front migration distance to be optimized before attempting the separation. A significant number of isocratic experiments are required to establish the relationship between solvent composition and analyte migration distance. Agreement of simulated separations with experiment are likely to be poor when the relationship between mobile phase composition and migration distance is not well determined (errors tend to accumulate through each step in the program). The predictive accuracy is also impaired if sorbent properties are modified by the solvents used in the previous development and by solvent demixing.

Optimized gradients for automated multiple development are usually arrived at by pragmatic means rather than computer simulation [9,61,67,104,178,179]. Two general approaches have been adopted for guided trial and error procedures. The first is based on the use of a universal gradient, which commences with methanol, ends with hexane, and uses dichloromethane or methyl *tert*-butyl ether as the intermediate or base solvent (typically employing 25 steps) [104]. By scaling and superimposing the chromatogram of the separation above the theoretical gradient profile, those regions of the chromatogram affecting the separation are easily identified. The appropriate initial mobile phase composition for the first development and the last development are easily identified, thus eliminating steps in the program that are not contributing to the separation. The gradient shape can then be modified to enhance resolution in those regions of the chromatogram that are poorly separated or to make better use of the zone capacity by minimizing regions devoid of sample zones. Thus, for example, the program can be modified to provide a shallower gradient over those regions of the chromatogram where peak separation is inadequate and steeper gradients in regions where the separated zones are well displaced from each other. For relatively simple mixtures this approach is often satisfactory. The universal gradient is not a linear solvent strength gradient, however, and the abrupt changes in solvent strength during the gradient can result in poor resolution by crowding sample components into a limited space [180].

Resolution may be inadequate in automated multiple development because of zone distortion, particularly tailing. This can be controlled by adjusting the layer conditioning step, or adding a tailing inhibitor in low concentration (e.g. formic acid, water, ammonia, etc.) to those compositions that are responsible for migration of the distorted zones. However, in those cases where the resolution remains inadequate after making the above adjustments, it is necessary to identify a different gradient composition for the separation. The Prisma model (section 6.8.1) can be used to identify more selective solvents to replace the initial, terminal, or base solvent; or for use in those program segments that affect the separation of poorly resolved zones in the universal gradient.

Alternatively, if the composition of the sample is known, and standards are available, isocratic plots of $R_M$ (section 6.3.1) against mobile phase composition can be used to infer suitable gradient separation conditions [181,182]. The initial solvent is chosen such that it possesses sufficient strength to cause migration of the most retained components of the mixture. The final solvent is selected to provide an acceptable separation of the least retained components without migrating them too close to the solvent front. The base solvent is selected based on its ability to provide optimum zone spacing throughout the chromatogram.

## 6.9 DETECTION

At best, inspection by eye of a thin-layer plate is capable of detecting about 1-10 micrograms of colored substances with reproducibility rarely better than 10-30%. Excising separated zones, eluting the substance from the sorbent, and determining the analyte concentration in solution by spectrophotometry is time consuming and often inaccurate. Difficulties in accurately locating the spot boundary by eye, incomplete elution of the sample from the sorbent and non-specific background absorption due to colloidal particles in the analytical solution are the main reasons for these deficiencies. Instruments for *in situ* recording and quantification of thin-layer chromatograms first appeared in the mid-1960s and are now considered essential for routine quantitative analysis in thin-layer chromatography [8,9].

All optical methods for the quantitative evaluation of thin-layer chromatograms are based on determining the difference in optical response between a sample free region of the layer and regions of the layer where sample components are present [9,11,183-186]. Transmission measurements can be used but reflectance is more common. In the reflectance mode most of the scattered light arises from particles close to the surface and is influenced less by variations in the layer thickness, which is responsible for much of the background noise in transmission measurements. Transmission measurements are limited to wavelengths greater than 320 nm due to strong absorption by the glass backing plate and by silica gel itself at shorter wavelengths. Reflectance measurements can be made at any wavelength from the UV to the near infrared (185-2500 nm).

When monochromatic light falls on an opaque medium some light is reflected from the surface, some absorbed by the medium and dissipated in some way, such as by conversion to heat, and the remainder is diffusely reflected or transmitted by the medium. For quantification it is the diffusely reflected/transmitted light that is of importance. The specularly reflected component only contributing to detector noise and not to the signal. The propagation of light within an opaque medium is a complex process that can only be solved mathematically if simplifying assumptions are made [9,11,184,186-189]. The most generally accepted theory is due to P. Kubelka and F. Munk, who proposed several solutions, the simplest being

$$(1 - R_{\infty})^2 / 2 R_{\infty} = 2.3 (aC / S) \qquad (6.14)$$

where $R_{\infty}$ is the reflectance for an infinitely thick opaque layer, a the molar absorption coefficient of the sample, C the sample concentration, and S the scatter coefficient for the layer. Equation (6.14) is derived explicitly for a layer of infinite thickness, which is not an accurate representation of separation layers, and this equation can only provide a phenomenological model for absorption measurements. However, it serves to illustrate the general properties of a solid-sorbent matrix on the observed sample response, and in a simple way, explains why calibration curves obtained on thin-layer plates do not obey Beer-Lambert's Law. It predicts, for example, a nonlinear relationship between signal and sample concentration in the absorption mode, an increase in response with larger molar absorption coefficients, and an increase in response for sorbents having a low scatter coefficient.

The absence of a suitable model for the relationship between absorption and sample concentration results in calibration being the principal method for quantification in thin-layer chromatography. Calibration curves are individual in shape, often pseudolinear at low sample concentrations curving towards the concentration axis at higher concentrations, and eventually reaching an asymptotic value where signal and sample concentration are no longer correlated, Figure 6.19. The pseudolinear region may be sufficiently wide for calibration for some compounds, but more generally nonlinear calibration is used based on a second order polynomial fit for calibration standards spanning the range of sample concentrations. Samples are quantified by interpolation only. A number of linearization methods have been proposed for calibration, but these

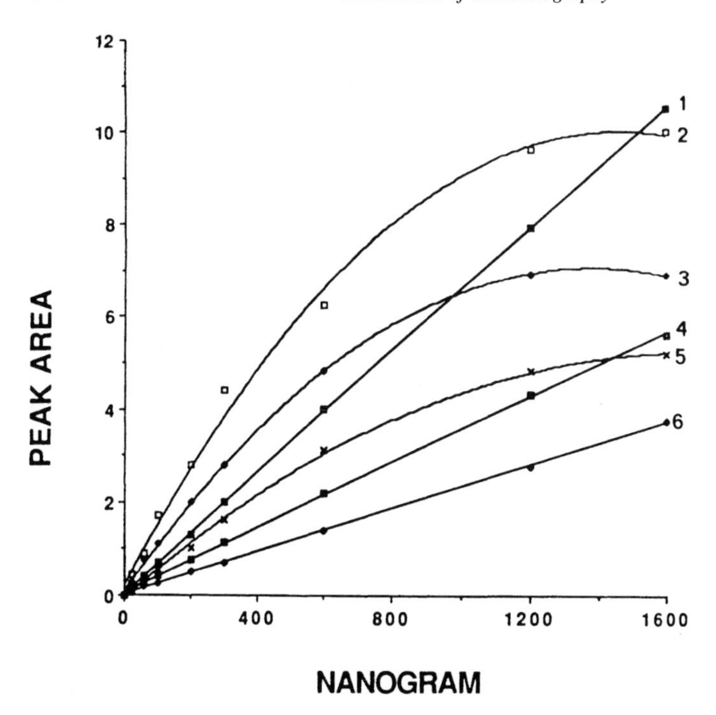


Figure 6.19. Typical calibration curves for different substances measured by absorption in the reflectance mode. Identification: 1 = practolol; 2 = azobenzene; 3 = diphenylacetylene; 4 = alprenolol; 5 = estrone; and 6 = pamatolol. (From ref. [173]; ©Elsevier).

seem both unnecessary and inappropriate, since they treat incorrectly the transposition of random errors in the data [183,184].

The fluorescence quenching technique is used to visualize sample zones that absorb UV-light on thin-layer plates containing a fluorescent indicator compound. When such a plate is exposed to UV light, UV-absorbing zones appear dark against a brightly fluorescing background of a lighter color. The UV-absorbing sample zones behave like an optical filter, absorbing a portion of the fluorescence excitation energy, which consequently diminishes the fluorescence emission intensity originating from the sample zone. The fluorescence quenching method is only applicable to substances that have an absorption spectrum that overlaps the excitation spectrum of the fluorescent indicator. Fluorescent indicators in common use have an absorption maximum around 280 nm and virtually no absorption below 240 nm. Fluorescence quenching is a less specific and less sensitive method of quantification than absorption measurements. It's main use is as a visualization technique for qualitative analysis.

Fluorescence measurements are fundamentally different to absorption measurements on thin-layer plates [184,190]. The fluorescence emission at low sample concentrations, F, is adequately described by $F = \phi I_0 abC$ where $\phi$ is the quantum yield, $I_0$ the intensity of the excitation source, a the molar absorption coefficient, b the layer thickness, and C the sample amount. Except for the sample amount, all terms in this expression are constant, or fixed by the experiment. Consequently, the fluorescence emission is linearly related to the sample amount. Calibration curves in fluorescence are usually linear over two or three orders of magnitude. For large sample sizes, self-absorption becomes a problem, and the fluorescence signal curves towards the concentration axis. The two-point calibration method provides an alternative calibration procedure that sustains sample throughput in screening studies when several components require quantification [191,192]. This method requires only a single standard concentration for each substance, and only a single lane for all calibration standards. The method is based on the linear relationship between slit width and fluorescence intensity in slit-scanning densitometry, and the linear relationship between this slope and sample amount. Thus, in practice, scanning each sample and the lane containing standards with two slit widths (e.g. 0.4 and 0.8 mm) is all that is required for calibration. The two-point calibration method is not a replacement for standard calibration procedures when the highest accuracy is required. It is used primarily to determine approximate sample amounts in large-scale screening programs.

### 6.9.1 Derivatization Techniques

There is a long history of using chemical reactions for the visualization of colorless compounds in thin-layer chromatography [5,6,11,193-195]. A primary reasons for employing thin-layer chromatography in analysis is the ease and flexibility of applying selective chemical reactions for the quantification of substances with poor UV-absorption characteristics and to obtain additional selectivity for functional group and family type detection. Different color-forming reactions combined with the measurement of migration distance in separation systems with complementary selectivity are widely used in forensic and environmental surveillance programs for the identification of drugs of abuse, pesticides, preservatives, etc. [13-15,196]. The combined information from chemically-selective reactions and selective chromatographic separations results in more certain identification of an unknown residue from many hundreds or thousands of possibilities in less time than the use of chromatographic techniques alone.

Prechromatographic reactions are favored when it is desired: to modify the properties of the sample to improve stability (i.e. minimize oxidative and catalytic degradations during measurements); to optimize chromatographic resolution; or to simplify the optimization of the reaction conditions [101,190,193]. On the other hand excess reagent, reaction by-products, catalysts used for the reaction, etc., may interfere with the detection of sample components because of their location and/or intensity in the chromatogram. In addition, some mixtures may be more difficult to separate because structural differences are minimized.

The advantages of postchromatographic reaction methods are that the separation is unaffected by the reaction, by-products of the reaction do not interfere in the chromatogram, and all samples and standards are derivatized simultaneously and with the same conditions. The reagent and derivative must have different detection characteristics, however, and the reagent must be evenly distributed over the layer. Since both prechromatographic and postchromatographic methods enhance the sensitivity and selectivity of the detection process, choosing between them will usually depend on the chemistry involved, ease of optimization, and which method best overcomes matrix and reagent interference.

For postchromatographic reactions the reagents can be applied to the layer through the gas phase or by evenly coating the layer with a solution of the reagents. Gas phase methods are fast and convenient but restricted by the number of useful reagents. Examples include iodine, ammonia, hydrogen chloride, etc., which are applied by inserting the layer into a tank containing a saturated atmosphere of the reactive vapor. Fluorescence can be induced in a number of compounds by heating them in a sealed chamber containing an activating reagent such as ammonium bicarbonate or zirconly sulfate [197]. The mechanism of this reaction remains a mystery and results vary widely for different compounds, but for some applications, it is useful.

Spraying, dipping or contact with a sponge soaked in reagent solution is used to apply reactive compounds in solution to the layer [194,198,199]. Spray techniques using simple atomizers have a long history of use in thin-layer chromatography, but reagent application by this method is difficult to perform well. The homogeneity of the reagent distribution over the layer depends on many factors such as droplet size, distance between the spray device and layer, direction of spraying, and discharge rate of the reagent. Manual spraying is not easy to control and if ventilation of the workspace is inadequate, it can be a health hazard. For quantitative methods, controlled immersion of the layer into a solution of the reagents is the preferred technique, since it does not rely on manual dexterity and produces superior results in scanning densitometry. Some solutions do not make good dipping solutions because they contain solvents that are too aggressive or viscous for convenient application (e.g. aqueous strong acids and bases). Dipping solutions are usually less concentrated than spray solutions and water is often replaced by an alcohol for adequate permeation of reversed-phase layers. In general, it is necessary to reformulate dipping solutions from earlier recipes for spray solutions, as well as possibly making changes to the reaction conditions [193,200]. Automated low volume dipping chambers are now available and are preferred for obtaining a uniform speed and dwell time for the immersion process. Immersion typically requires only a few seconds, which is long enough to impregnate the layer with solution but not long enough to wash sample components off the layer. In the absence of automation, dipping techniques share some of the problems inherent with spraying, in that the results depend on manual dexterity.

Postchromatographic derivatization reactions can be classified as reversible or destructive depending on the type of interaction the reagents undergo with the separated compounds, and as selective or universal, based on the specificity of the reaction. The most common reversible methods employ iodine vapor, water, fluorescein, or pH indicators as visualizing reagents [193-95,201]. In the iodine vapor method, the dried plate is enclosed in a chamber containing a few crystals of iodine; substance zones are stained more rapidly than the layer and appear as yellow-brown spots on a light yellow background. As little as 0.1 to 0.01 μg of sample can be visualized in this way. Simply removing the plate from the visualization chamber and allowing the iodine to evaporate reverses the reaction. Spraying the layer with water reveals hydrophobic compounds as white spots on a translucent background if the water-moistened layer is viewed against a white light. Solutions of pH indicators (e.g. bromocresol green, bromophenol blue, etc.) are widely used for the detection of acidic and basic compounds. The above methods are all fairly universal and reversible, and can be used for detection when the sample is required for further studies. Occasionally these methods are also used for quantitative analysis.

Irreversible methods are common for quantification and comprise hundreds of reagents based on selective chemistries reduced to standard operations over several decades of use [11,193]. A typical example is the use of copper sulfate or manganese chloride in phosphoric acid solution for the quantitative analysis of lipids. After spraying or dipping in the acid solution and heating, the lipids are converted to colored zones with a nearly uniform response [202]. Functional group specific or compound class selective reagents are used for the determination of low levels of substances in complex matrices, such as biological fluids and plant extracts.

### 6.9.2 Sorbent-Aided Response Modification

The fluorescence emission for substances in thin-layer chromatography are at times less than expected when predicted from solution experiments, occur at different excitation and emission wavelengths compared with solution spectra, and may produce a steadily decaying signal [101,184,203]. Adsorption on the layer provides additional nonradiative pathways for the dissipation of the excitation energy, which is most likely lost as heat to the surroundings, reducing the fluorescence signal. Quenching of the signal by interaction with oxygen, or reaction of the solute with oxygen to produce new products with a diminished fluorescence yield, are other known mechanisms affecting the magnitude and stability of the fluorescence response.

The extent of fluorescence quenching often depends on the layer type and is usually more significant for silica gel than for chemically bonded layers. In some cases, the emission signal can be enhanced if the layer is impregnated with a viscous liquid before scanning. Common fluorescence-enhancing reagents include solutions of liquid paraffin, Triton X-100 and Fomblin oils [101,184,204]. The general mechanism of fluorescence enhancement is assumed to involve (partial) dissolution of the adsorbed analyte with an increase in emission resulting from the fraction of analyte transferred to the liquid phase where fluorescence quenching is less severe. In favorable cases, a signal enhancement of 10- to 200-fold is obtained. In those cases where fluorescence quenching is negligible, the application of a fluorescence-enhancing reagent rarely produces a significant increase in emission. Solvents of high viscosity are employed as fluorescence enhancing reagents to minimize zone broadening due to diffusion of the dissolved analyte during the measurement process.

Oxygen is a ubiquitous fluorescence-quenching reagent and is difficult to exclude from the layer surface during scanning densitometry. Flooding the scanning stage with nitrogen is possible. It is generally more practical, however, to impregnate the layer with an antioxidant, such as BHT (2,6-di-*tert*-butyl-4-methylphenol) before sample application, and to add antioxidant to the mobile phase [101,184,205]. Some adsorbed samples undergo photoxidation reactions, requiring the presence of an antioxidant and precautions to shield the sample from light. Fortunately, the above problems are not severe in most cases, and only those substance that are unusually susceptible to catalytic oxidation or oxygen quenching require special handling.

### 6.9.3 Slit-Scanning Densitometers

Instruments for scanning densitometry share many features in common, Figure 6.20 [8,9,11,104,140,184,204]. They usually allow measurements in the reflectance or transmission mode, by absorbance or fluorescence. Halogen or tungsten lamps are used for the visible range and deuterium lamps for the UV region. High-intensity mercury or xenon arc lamps are used for fluorescence measurements. In most cases, a motorized grating monochromators is used for wavelength selection and to record absorption spectra. For fluorescence measurements a filter, which transmits the emission wavelength envelope but attenuates the excitation wavelength, is placed between the detector and the plate. Photomultipliers or photodiodes are generally used for signal measurements. The plate is scanned at speeds up to about 10 cm/s by mounting it on a movable stage controlled by stepping motors. For lane scanning a fixed sample beam is shaped into a rectangular area on the plate surface through which the plate is transported. Some instruments have a turntable-type scanning stage for peripheral or radial scanning of circular and anticircular chromatograms. Alternatively, for point scanning the measuring beam is shaped into a spot or rectangle with dimensions much smaller than the sample zones. By moving the scanning stage in the x and y direction a zigzag or meander scan is initiated. Zigzag and meander scanning allow zones of any shape to be quantified. Distorted separations resulting from lateral migration of sample components can be corrected by track optimization in which the sample zones are integrated as if the slit had moved along an optimum track from peak maximum to peak maximum. Dual-wavelength measurements can be used to minimize matrix interference if suitable wavelengths can be found for the measurements. Ideally, two similar wavelengths are required at which sample absorption occurs at only one wavelength. Since absorption spectra tend to be broad this criterion is not easy to meet in practice.

A laboratory-built scanning densitometer using a fiber optic bundle for illumination of the layer and collection of reflected light (or fluorescence) in conjunction with a photodiode array detector affords some attractive features that conventional scanning instruments lack [186,206,207]. These include: the simultaneous recording of spectra from 198-612 nm with a spectral resolution of 0.8 nm; simultaneous detection at multiple wavelengths; a spatial resolution $< 0.16$ mm; and implementation of a wide

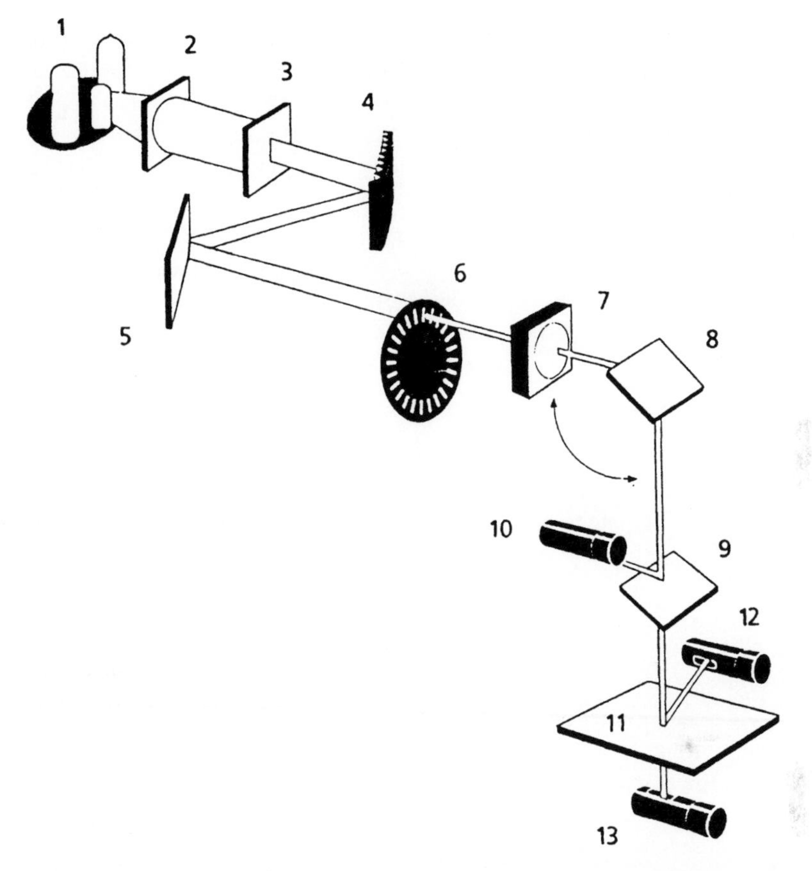


Figure 6.20. Optical arrangement for a slit-scanning densitometer (Camag scanner 3). Identification: 1 = lamp selector; 2 = entrance lens slit; 3 = monochromator entry slit; 4 = grating; 5 = mirror; 6 = slit aperture disk; 7 = lens system; 8 = mirror; 9 = beam splitter; 10 = reference photomultiplier; 11 = TLC plate; 12 = measuring photomultiplier; and 13 = photodiode for transmission measurements. (From ref. [8]; ©Marcel Dekker)

range of chemometic procedures for library searches, purity evaluation, resolution of overlapping zones, etc.

The principal sources of error in scanning densitometry are the reproducibility of sample application, the reproducibility of chromatographic conditions, the reproducibility of positioning the spot in the center of the measuring beam, and the reproducibility of the measurement [31,208,209]. The measurement error can be determined by repeatedly scanning a single lane of the thin-layer plate without changing any experimental variables between scans. It is composed of errors due to the optical measurement, electronic amplification, and the recording device. The measurement error is dependent on the signal-to-noise ratio, but for a properly adjusted instrument, typical relative standard deviation values fall into the range 0.2-0.7%. The positioning error is negligible if band application or meander scanning is used and can be made very small in linear scanning of spots if good analytical practices are followed. The sample application error can be determined by applying standards to the layer and scanning the applied zones before development without changing any experimental variables between scans. The relative standard deviation for application errors using automated sample applicators is about 1.0 %. If the plate is then developed and rescanned without changing any of the experimental variables the chromatographic error can be estimated. This is quite often the most significant error and is only reduced by minimizing variations in the development process. The relative standard deviation from all errors in scanning densitometry can be maintained below 2-3 % for high performance layers.

### 6.9.4 Video Densitometers

For video densitometry optical scanning takes place electronically, using a computer with video digitizer, light source, monochromators and appropriate optics to illuminate the plate and focus the image onto a charged-coupled device (CCD) video camera [11,103,210-213]. The plate is evenly illuminated with monochromatic light and the reflected light focused as a scaled image of the plate directly onto the active element of the CCD camera. The CCD camera functions as a two-dimensional array of unit detectors (pixels). Photons colliding with an individual pixel result in the creation of a charge stored by an applied voltage. Accumulated charges on each pixel are then transported to a readout amplifier using sequential clocking of the applied voltage and the signal digitized for computer analysis. The captured images are initialized, stored, and transformed by the computer into chromatographic data. Background subtraction and thresholding are common data transformation processes. For background subtraction the accumulated images of a blank plate are subtracted from the analytical plate on a pixel-by-pixel basis. Thresholding is used to prevent the occurrence of negative values for the plate luminescence.

The main attractions of video densitometry for detection in thin-layer chromatography are fast data acquisition, absence of moving parts, simple instrument design, and compatibility with data analysis of two-dimensional chromatograms that are difficult to scan using conventional slit-scanning densitometers. However, given the limitation of today's technology, video densitometers cannot compete with mechanical scanners in terms of sensitivity, resolution, and available wavelength measuring range, but these instruments continue to improve all the time. They have proven popular in the development of field-portable instruments and as a replacement for photographic documentation of thin-layer separations. Modern instruments provide attractive options for searching and comparing sample images as well as integration of peak areas [214,215]. Most low-cost instruments are designed for operation in the visible region or UV region when layers containing a fluorescence indicator are used. Surprisingly good results were obtained with a conventional office flat bed scanner modified for scanning thin-layer plates by inclusion of a light source [216,217].

### 6.9.5 Miscellaneous Methods

Detection in thin-layer chromatography is dominated by optical methods using scanning densitometry with video densitometry as an increasingly popular alternative. Both approaches, however, yield similar information. Other, alternative methods, are useful because they provide complementary information, even if used less frequently. Identification methods based on mass spectrometry (section 9.2.2.5) and infrared and Raman spectroscopy (section 9.3.1.4) and fluorescence line narrowing spectroscopy are discussed elsewhere [11,218,219]. Flame ionization has been used for the detection of samples lacking a chromophore for optical detection. The separation is performed on specially prepared thin, quartz rods with a surface coating of adsorbent attached to the rod by sintering [80,81,202,220]. The rods are developed in the normal way, usually held in a support frame that also serves as the scan stage after the rods have been removed from the developing chamber and dried. Several rods can be held in the support frame and automatically scanned in order. The rods are moved at a controlled speed through a hydrogen flame and the signal processed in a similar manner to the flame ionization detector used in gas chromatography. The linear working range of the detector is about 3-30 $\mu$g for most substances and the response nearly universal. Other detector options include the thermionic ionization and flame photometric detectors for element-selective detection.

Biologically active compounds separated by thin-layer chromatography are easily revealed by bioautographic detection (biomonitoring) [11]. The thin-layer plate is simply dipped or sprayed with an enzyme such as cholinesterase and a substrate to determine inhibition by color differentiation [221,222]. Alternatively, the plate can be dipped into a suspension of luminescent bacteria with toxic substances revealed as dark zones resulting from reduced bioluminescence [223-225]. Antibacterial compounds can be detected by contacting the layer face-to-face with an inoculated agar plate. The separated zones are transferred to the agar gel by diffusion and inhibition zones visualized by use of suitable stains and (mainly) dehydrogenase-activity-detecting reagents [11,226]. Thin-layer chromatography with biomonitoring is a useful technique for toxicity-directed wastewater analysis [224]. This approach eliminates the time consuming work-up of fractions for biological testing and substantially reduces the expenditure for testing itself. To avoid the time consuming isolation of already known active compounds, a dereplication step (recognizing the already known active compounds) is required for the isolation of novel biologically active agents in screening plant and animal sources. Thin-layer chromatography with biomonitoring is well suited to this purpose, and allows unique compounds to be identified for isolation from complex extracts. Thin-layer chromatography with biomonitoring was also shown to be useful for high throughput screening of combinatorial libraries [226].

Modern imaging detectors for radiolabeled compounds employ a range of different technologies for the position sensitive detection of $\beta$- and $\gamma$-emitting isotopes of (mainly) $^{3}H$, $^{14}C$, $^{32}P$ and $^{125}I$ [227-233]. These instruments are only likely to be found in laboratories that handle radiolabeled compounds on a routine basis and have the necessary safety and training protocols in place. For detection multiwire proportional counters, microchannel array detectors or phosphor imaging analyzers are used. Most detectors provide a spatial resolution of < 1 mm and are able to detect zones of radioactivity on a plate containing less than 10 dpm in a relatively short exposure time. These computer-based systems offer a variety of methods for data analysis and display.

## 6.10 REFERENCES


[1] C. F. Poole, J. Chromatogr. A 856 (1999) 399.
[2] F. Kreuzig, J. Planar Chromatogr. 11 (1998) 322.
[3] V. Berezkin, J. Planar Chromatogr. 8 (1995) 401.
[4] L. S. Ettre, Chromatographia 51 (2000) 7.
[5] J. G. Kirchner, *Thin-Layer Chromatography*, Wiley, New York, 1978.
[6] E. Stahl, Thin-Layer Chromatography, Springer-Verlag, New York, 1969.
[7] A. Zlatkis and R. E. Kaiser (Eds.), *HPTLC High Performance Thin-Layer Chromatography*, Elsevier, Amsterdam, 1977.
[8] W. Bertsch, S. Hara, R. E. Kaiser and A. Zlatkis (Eds.), *Instrumental HPTLC*, Huthig, Heidelberg, 1980.
[9] J. Sherma and B. Fried (Eds.), *Handbook of Thin-Layer Chromatography*, Marcel Dekker, New York, 1996.
[10] C. F. Poole and S. K. Poole, Anal. Chem. 66 (1994) 27A.
[11] Sz. Nyiredy (Ed.), *Planar Chromatography. A Retrospective View for the Third Millennium*, Springer, Budapest, Hungary, 2001.
[12] B. Fried and J. Sherma (Eds.), *Practical Thin-Layer Chromatography. Multidisciplinary Approach*, CRC Press, Boca Raton, FL, 1996.
[13] I. Ojanpera, K. Goebel and E. Vuori, J. Liq. Chromatogr. & Rel. Technol. 22 (1999) 161.
[14] R. A. de Zeeuw, J. P. Frank, F. Degel, G. Machbert, H. Scutz and J. Wijsbeek (Eds.), *Thin-Layer Chromatographic $R_F$ Values of Toxicologically Relevant Substances on Standardized Systems*, VCH, Weinheim, 1992.
[15] T. Moffat, D. Osselton and B. Widdop (Eds.), *Clarke's Isolation and Identification of Drugs*, Royal Pharmaceutical Society, London, 2002.
[16] F. Smets, Ch. Vanhoenackere and G. Pottie, Anal. Chim. Acta 275 (1993) 147.
[17] J. Unruh, D. P. Schwartz, R. A. Barford, J. AOAC Int. 76 (1993) 335.
[18] J. P. Abjean, J. Planar Chromatogr. 6 (1993) 147.
[19] J. P. Abjean, J. AOAC Int. 80 (1997) 737.
[20] G. E. Morlock, J. Chromatogr. A 754 (1996) 423.
[21] S. Butz and H.-J. Stan, Anal. Chem. 67 (1995) 620.
[22] L. Gagliadi, D. De Orsi, M. R. Del Giudice, F. Gatta, R. Porra, P. Chimenti and D. Tonelli, Anal. Chim. Acta 457 (2002) 187.
[23] S. Zellmer and J. Lasch, J. Chromatogr. B 691 (1997) 321.
[24] J. Muthig and H. Ziehr, J. Chromatogr. B 687 (1996) 357.
[25] S. K. Poole, W. Kiridena, K. G. Miller and C. F. Poole, J. Planar Chromatogr. 8 (1995) 257.
[26] K. Hostettmann, C. Terreaux, A. Marston and O. Potterat, J. Planar Chromatogr. 10 (1997) 251
[27] H. Wagner, S. Bladt and E. M. Zgainski, *Plant Drug Analysis*, Springer-Verlag, Berlin, 1984.
[28] B. Renger, J. AOAC Int. 76 (1993) 7.
[29] B. Renger, J. AOAC Int. 84 (2001) 1217.
[30] K. Ferenczi-Fodor, Z. Vegh, A. Nagy-Turak, B. Renger and M. Zeller, J. AOAC Int. 84 (2001) 1265.
[31] G. Szepesi and Sz. Nyiredy, J. Pharm. Biomed. Anal. 10 (1992) 1007.
[32] F. Geiss, *Fundamentals of Thin Layer Chromatography*, Huethig, Heidelberg, 1987.

[33] G. Guiochon and A. Siouffi, J. Chromatogr. Sci. 16 (1978) 598.
[34] C. F. Poole, J. Planar Chromatogr. 2 (1989) 95.
[35] L. S. Litvinova and O. I. Kurenbin, J. Planar Chromatogr. 4 (1991) 402.
[36] S. Nyiredy, Z. Fater, L. Botz and O. Sticher, J. Planar Chromatogr. 5 (1992) 308.
[37] K. Dross, C. Sonntag and R. Mannhold, J. Chromatogr. 639 (1993) 287.
[38] W. P. N. Fernando and C. F. Poole, J. Planar Chromatogr. 3 (1990) 389.
[39] G. Guiochon, G. Korosi and A. Siouffi, J. Chromatogr. Sci. 18 (1980) 324.
[40] P. Merkku, J. Yliruusi and H. Vuorela, J. Planar Chromatogr. 8 (1995) 112.
[41] Sz. Nyiredy and G. Szepesi, J. Pharm. Biomed. Anal. 10 (1992) 1017.
[42] Sz. Nyiredy, L. Botz and O. Sticher, J. Planar Chromatogr. 2 (1989) 53.
[43] R. E. Kaiser and R. I. Reider, J. Am. Oil Chem. Soc. 66 (1989) 79.
[44] G. Flodberg and J. Roeraade, J. Planar Chromatogr. 8 (1995) 10.
[45] D. Nurok, Anal. Chem. 72 (2000) 634A.
[46] S. Nyiredy, Trends Anal. Chem. 20 (2001) 91.
[47] Sz. Nyiredy, S. Y. Meszaros, K. Dallenbach-Tolke, K. Nyiredy-Mikita, and O. Sticher, J. High Resolut. Chromatogr. 10 (1987) 352.
[48] A. Velayudhan, B. Lillig and C. Horvath, J. Chromatogr. 435 (1988) 397.
[49] Sz. Nyiredy and Z. Fater, J. Planar Chromatogr. 7 (1994) 329.
[50] C. F. Poole and I. D. Wilson, J. Planar Chromatogr. 10 (1997) 332.
[51] T. Shafik, A. G. Howard, F. Moffatt and I. D. Wilson, J. Chromatogr. A 841 (1999) 127.
[52] I. Malinowska, J. Planar Chromatogr. 13 (2000) 4.
[53] D. Nurok, M. C. Frost and D. M. Chenoweth, J. Chromatogr. A 903 (2000) 211.
[54] D. Nurok, J. M. Koers, D. A. Nyman and W. Liao, J. Planar Chromatogr. 14 (2001) 409.
[55] G. Guiochon and A. Siouffi, J. Chromatogr. Sci. 16 (1978) 470.
[56] A. Siouffi, F. Bressolle and G. Guiochon, J. Chromatogr. 209 (1981) 129.
[57] C. F. Poole, J. Planar Chromatogr. 1 (1988) 373.
[58] B. Belenkii, O. Kurenbin, L. Litvinova and E. Gankina, J. Planar Chromatogr. 3 (1990) 340.
[59] C. F. Poole and W. P. N. Fernando, J. Planar Chromatogr. 5 (1992) 323.
[60] J. Bladek and A. Rostkowski, J. Planar Chromatogr. 10 (1997) 163.
[61] C. F. Poole and S. K. Poole, J. Chromatogr. A 703 (1995) 573.
[62] W. P. N. Fernando and C. F. Poole, J. Planar Chromatogr. 4 (1991) 278.
[63] L. Botz, Sz. Nyiredy and O. Sticher, J. Planar Chromatogr. 4 (1991) 115.
[64] E. Tyihak, G. Katay, Z. Ostorics and E. Mincsovics, J. Planar Chromatogr. 11 (1998) 5.
[65] W. P. N. Fernando and C. F. Poole, J. Planar Chromatogr. 5 (1992) 50.
[66] C. F. Poole and W. P. N. Fernando, J. Planar Chromatogr. 6 (1993) 357.
[67] C. F. Poole and M. T. Belay, J. Planar Chromatogr. 4 (1991) 345.
[68] C. F. Poole, S. K. Poole and M. T. Belay, J. Planar Chromatogr. 6 (1993) 438
[69] B. Szabady, M. Ruszinko and Sz. Nyiredy, J. Planar Chromatogr. 8 (1995) 279.
[70] B. Szabady, M. Ruszinko and Sz. Nyiredy, Chromatographia 45 (1997) 369.
[71] C. F. Poole, S. K. Poole, W. P. N. Fernando, T. A. Dean, H. D. Ahmed and J. A. Berndt, J. Planar Chromatogr. 2 (1989) 336.
[72] C. F. Poole and S. K. Poole, J. Planar Chromatogr. 2 (1989) 165.
[73] S. Essig and K.-A. Kovar, J. Planar Chromatogr. 10 (1997) 114.
[74] B. Klama and T. Kowalska, J. Planar Chromatogr. 10 (1997) 427.
[75] G. Guiochon and A. Siouffi, J. Chromatogr. 245 (1982) 1.
[76] E. Hahn-Deinstrop, J. Planar Chromatogr. 5 (1992) 57.
[77] H. Halpaap and K.-F. Krebs, J. Chromatogr. 142 (1977) 823.
[78] G. K. Bauer, A. M. Pfeifer, H. E. Hauck and K.-A. Kovar, J. Planar Chromatogr. 11 (1998) 94.
[79] H. E. Hauck, O. Bund, W. Fischer and M. Schulz, J. Planar Chromatogr. 14 (2001) 234.
[80] M. Ranny, Thin-Layer Chromatography with Flame Ionization Detection, D. Reidel Publishing Co., Dordrecht, Netherlands, 1987.

[81] R. G. Ackman, C. A. Macleod and A. K. Banerjee, J. Planar Chromatogr. 3 (1990) 452.
[82] K. K. Unger (Ed.), *Packings and Stationary Phases in Chromatographic Techniques*, Marcel Dekker, New York, 1990, p. 251-330.
[83] D. L. Grumprecht, J. Chromatogr. 595 (1992) 368.
[84] J. A. Berndt and C. F. Poole, J. Planar Chromatogr. 1 (1988) 174.
[85] H. E. Hauck and H. Halpaap, Chromatographia 13 (1980) 538.
[86] S. K. Poole, H. D. Ahmed, M. T. Belay, W. P. N. Fernando and C. F. Poole, J. Planar Chromatogr. 3 (1990) 133.
[87] H. Halpaap and J. Ripphahn, Chromatographia 10 (1977) 643.
[88] W. Funk, G. Donnevert, B. Schuch, V. Gluck and J. Becker, J. Planar Chromatogr. 2(1989) 317.
[89] T.-S. Li, J.-T. Li and H.-Z. Li, J. Chromatogr. A 715 (1995) 372.
[90] J. Ahmad, J. Planar Chromatogr. 9 (1996) 236.
[91] M. Waksmundzka-Hajnos, Chromatographia 43 (1996) 640.
[92] U. A. Th. Brinkman, Trends Anal. Chem. 5 (1986) 178.
[93] W. Jost and H. E. Hauck, Adv. Chromatogr. 27 (1987) 129.
[94] E. Heilweil and F. Rabel, J. Chromatogr. Sci. 23 (1985) 101.
[95] W. Kiridena and C. F. Poole, J. Planar Chromatogr. 12 (1999) 13.
[96] G. Kowalik and T. Kowalska, J. Planar Chromatogr. 14 (2001) 224.
[97] W. Jost and H. E. Hauck, J. Chromatogr. 261 (1983) 235.
[98] W. Jost, H. E. Hauck, Anal. Biochem. 135 (1983) 120.
[99] H. E. Hauck, J. Planar Chromatogr. 8 (1995) 346.
[100] E. Hahn-Deinstrop, J. Planar Chromatogr. 6 (1993) 313.
[101] C. F. Poole, S. K. Poole, T. A. Dean and N. M. Chirco, J. Planar Chromatogr. 2 (1989) 180.
[102] R. J. Maxwell, A. R. Lightfield, J. Planar Chromatogr. 12 (1999) 109.
[103] Y. Liang, M. E. Baker, D. A. Gilmore and M. Bonner Denton, J. Planar Chromatogr. 9 (1998) 247.
[104] D. E. Jaenchen and H. J. Issaq, J. Liquid Chromatogr. 11 (1988) 1941.
[105] J. K. Rozylo, V. G. Berezkin, I. Malinowska and A. Jamrozek-Manko, J. Planar Chromatogr. 14 (2001) 272.
[106] C. W. Maboundou, P.-Y. Grosse, P. Delvordre and N. Vermerie, J. Planar Chromatogr. 12 (1999) 373.
[107] D. C. Fenimore and C. J. Meyer, J. Chromatogr. 186 (1979) 555.
[108] G. Malikin, S. Lam and A. Karman, Chromatographia 18 (1984) 253.
[109] E. Stahl, J. Chromatogr. 142 (1977) 15.
[110] E. Stahl and W. Schild, Fresenius' Z. Anal. Chem. 280 (1976) 99.
[111] U. Keller and I. Flament, Chromatographia 28 (1989) 445.
[112] L. Wunsche, U. Keller and I. Flament, J. Chromatogr. 552 (1991) 539.
[113] G. L. De Vault and M. J. Sepaniak, J. Microcol. Sep. 12 (2000) 419.
[114] G. W. Somsen, C. Gooijer and U. A. Th. Brinkman, Trends Anal. Chem. 17 (1998) 129.
[115] D.E. Jaenchen and H.J. Issaq, J. Liq. Chromatogr. 11 (1988) 1941.
[116] K. Burger, Analusis 18 (1990) i113.
[117] C.T. Banks, J. Pharm. Biomed. Anal. 11 (1993) 705.
[118] O.R. Queckenberg and A.W. Frahn, J. Planar Chromatogr. 6 (1993) 55.
[119] H.-J. Stan and F. Schwarzer, J. Chromatogr. A 819 (1998) 35.
[120] M. A. Hawryl and E. Soczewinski, Chromatographia 52 (2000) 175.
[121] E. Muller and H. Jork, J. Planar Chromatogr. 6 (1993) 21.
[122] D. C. Fenimore and C. M. Davis, Anal. Chem. 53 (1981) 252A.
[123] R. E. Kaiser, J. Planar Chromatogr. 1 (1988) 265.
[124] R. E. Kaiser (Ed.), *Planar Chromatography*, Huthig, Heidelberg, vol. 1, 1986.
[125] J. A. Perry, J. Chromatogr. 165 (1979) 117.
[126] R. E. Tecklenburg, G. H. Fricke and D. Nurok, J. Chromatogr. 290 (1984) 75.
[127] H. Cortes (Ed.), *Multidimensional Chromatography. Techniques and Applications*, Marcel Dekker, New York, 1990.

[128] S. K. Poole and C. F. Poole, J. Planar Chromatogr. 5 (1992) 221.
[129] T. H. Dzido, M. A. Hawryl, W. Golkiewicz and E. Soczewinski, J. Planar Chromatogr. 8 (1995) 306.
[130] M. T. Belay and C. F. Poole, J. Planar Chromatogr. 4 (1991) 424.
[131] S. A. Schuette and C. F. Poole, J. Chromatogr. 239 (1982) 251.
[132] M. Zakaria, M.-F. Gonnord and G. Guiochon, J. Chromatogr. 271 (1983) 127.
[133] H. J. Issaq, Trends Anal. Chem. 9 (1990) 36.
[134] G. Guiochon, L. A. Beaver, M. F. Gonnord, A.-M. Siouffi and M. Zakaria, J. Chromatogr. 255 (1983) 415.
[135] M.-F. Gonnord and A.-M. Siouffi, J. Planar Chromatogr. 3 (1990) 206.
[136] M. Bathori, G. Blunden and H. Kalasz, Chromatographia 52 (2000) 815.
[137] E. Soczewinski, M. A. Hawryl and A. Hawryl, Chromatographia 54 (2001) 789.
[138] M. A. Hawryl, A. Hawryl and E. Soczewinski, J. Planar Chromatogr. 15 (2002) 4.
[139] M. Glensk, Z. Bialy, M. Jurzysta and W. Cisowski, Chromatographia 54 (2001) 669.
[140] H. Yamamoto, K. Nakamura, D. Nakatani and H. Terada, J. Chromatogr. 543 (1991) 201.
[141] M. Petrovic, M. Kostelan-Macan and S. Balic, J. Planar Chromatogr. 11 (1998) 353.
[142] P. Petrin, J. Chromatogr. 123 (1976) 65.
[143] T. H. Dzido, J. Planar Chromatogr. 6 (1993) 78.
[144] K. Burger, Analusis 18 (1990) i113.
[145] D. W. Armstrong, K. H. Bui and R. E. Boehm, J. Liq. Chromatogr. 6 (1983) 1.
[146] E. S. Gankina, I. I. Efinova, J. J. Kever and B. G. Belenkii, Talanta 34 (1987) 167.
[147] D. Nurok, Chem. Rev. 89 (1989) 363.
[148] A.-M. Siouffi, J. Chromatogr. 556 (1991) 81.
[149] J. K. Rozylo and R. Siembida, J. Planar Chromatogr. 10 (1997) 97.
[150] E. Reich and T. George, J. Planar Chromatogr. 10 (1997) 273.
[151] C. F. Poole and N. C. Dias, J. Chromatogr. A 892 (2000) 123.
[152] E. Hidvegi, S. Perneczki and M. Forstner, J. Planar Chromatogr. 13 (2000) 414.
[153] F. L. Birkenshaw and D. G. Waters, J. Planar Chromatogr. 8 (1995) 319.
[154] S. J. Costanzo, J. Chromatogr. Sci. 35 (1997) 156.
[155] I. Malinowska and J. K. Rozylo, J. Planar Chromatogr. 10 (1997) 411.
[156] Sz. Nyiredy, K. Dallenbach-Toelke and O. Sticher, J. Planar Chromatogr. 1 (1988) 336.
[157] Sz. Nyiredy and Z. Fater, J. Planar Chromatogr. 8 (1995) 341.
[158] A. Pelander, K. Sivonen, I. Ojanpera and H. Vuorela, J. Planar Chromatogr. 10 (1997) 434.
[159] A. Pelander, J. Summanen, T. Yrjonen, H. Haario, I. Ojanpera and H. Vuorela, J. Planar Chromatogr. 12 (1999) 365.
[160] P. Vuorela, E.-L. Rahko, R. Hiltunen and H. Vuorela, J. Chromatogr. A 670 (1994) 191.
[161] M. H. Abraham, C. F. Poole and S. K. Poole, J. Chromatogr. A 749 (1996) 201.
[162] W. Kiridena and C. F. Poole, J. Chromatogr. A, 802 (1998) 335.
[163] N. C. Dias and C. F. Poole, J. Planar Chromatogr. 13 (2000) 337.
[164] N. C. Dias and C. F. Poole, J. Planar Chromatogr. 14 (2001) 160.
[165] K. Morita, S. Koike and T. Aishima, J. Planar Chromatogr. 11 (1998) 94.
[166] Q.-S. Wang and B.-W. Yan, J. Planar Chromatogr. 9 (1996) 192.
[167] M. C. Frost, T. Lahr, R. M. Kleyle and D. Nurok, J. Chromatogr. A 788 (1997) 207.
[168] D. Nurok, R. M. Kleyle, C. L. McCain, D. S. Risley and K. J. Ruterbories, Anal. Chem. 69 (1997) 1398.
[169] W. Markowski and K. L. Czapinska, J. Liq. Chromatogr. 18 (1995) 1405.
[170] T. H. Dzido, G. Matysik and E. Soczewinski, J. Planar Chromatogr. 4 (1991) 161.
[171] G. Matysik and E. Soczewinski, J. Planar Chromatogr. 9 (1996) 404.
[172] G. Matysik, Chromatographia 43 (1996) 39.
[173] G. Matysik, Chromatographia 43 (1996) 301.
[174] E. Soczewinski and W. Markowski, J. Chromatogr. 370 (1986) 63.
[175] W. Markowski and E. Soczewinski, J. Chromatogr. 623 (1992) 139.

[176] W. Markowski, J. Chromatogr. 635 (1993) 283.
[177] W. Markowski, J. Chromatogr. 726 (1993) 185.
[178] J. Summanen, R. Hiltunen and H. Vuorela, J. Planar Chromatogr. 11 (1998) 16.
[179] Y. Scholl, N. Asano and B. Drager, J. Chromatogr. A 928 (2001) 217.
[180] P. V. Colthup, J. A. Bell and D. L. Gadsdon, J. Planar Chromatogr. 6 (1993) 386.
[181] G. Lodi, A. Betti, V. Brandolini, E. Menziani and B. Tosi, J. Planar Chromatogr. 7 (1994) 29.
[182] G. Lodi, A. Betti, E. Menziani, V. Brandolini and B. Tosi, J. Planar Chromatogr. 4 (1991) 106.
[183] S. Ebel, J. Planar Chromatogr. 9 (1996) 4.
[184] C. F. Poole and S. K. Poole, J. Chromatogr. 492 (1989) 539.
[185] V. A. Pollak, J. Planar Chromatogr. 6 (1993) 7.
[186] B. Spangenberg, P. Prost and S. Ebel, J. Planar Chromatogr. 15 (2002) 88.
[187] I. E. Bush and H. P. Greeley, Anal. Chem. 56 (1984) 91.
[188] F. A. Huf, J. Planar Chromatogr. 1 (1988) 46.
[189] I. Vovk, M. Franko, J. Gibkes, M. Prosek and D. Bicanic, J. Planar Chromatogr. 10 (1997) 258.
[190] W. R. G. Baeyens and B. L. Ling, J. Planar Chromatogr. 1 (1988) 198.
[191] H. T. Butler, M. E. Coddens, S. Khatib and C. F. Poole, J. Chromatogr. Sci. 23 (1985) 200.
[192] H. T. Butler and C. F. Poole, J. Chromatogr. Sci. 21 (1983) 385.
[193] H. Jork, W. Funk, W. Fischer and H. Wimmer, *Thin-Layer Chromatography. Reagents and Detection Methods*, VCH, Weinheim, vol. 1 (1990) and vol. 2 (1992).
[194] G. D. Barrett, Adv. Chromatogr. 11 (1974) 145.
[195] E. Hahn-Deinstrop, *Applied Thin-Layer Chromatography – Best Practice and Avoidance of Mistakes*, Wiley-VCH, Weinhein, Germany, 2000.
[196] T. Imrag and A. Junker-Buchhelt, J. Planar Chromatogr. 9 (1996) 39.
[197] R. J. Maxwell and J. Unruh, J. Planar Chromatogr. 5 (1992) 35.
[198] P. Delvordre, E. Postaire, C. Regnault and C. Sarbach, J. Planar Chromatogr. 3 (1990) 500.
[199] E. Hahn-Deinstrop, A. Koch and M. Muller, Chromatographia 51 (2000) S-302.
[200] M. J. Kurantz, R. J. Maxwell and M. Cygnarowicz-Provost, J. Planar Chromatogr. 5 (1992) 41.
[201] W. Wardas and A. Pyka, J. Planar Chromatogr. 14 (2001) 8.
[202] S. Zellmer and J. Lasch, J. Chromatogr. B 691 (1997) 321.
[203] F. P. Cossio, A. Arrieta, V. L. Cebolla, L. Membrado, M. P. Domingo, P. Henrion and J. Vela, Anal. Chem. 72 (2000) 1759.
[204] C. F. Poole, M. E. Coddens, H. T. Butler, S. A. Schuette, S. S. J. Ho, S. Khatib, L. Piet and K. K. Brown, J. Liq. Chromatogr. 8 (1985) 2875.
[205] M. J. Cikalo, S. K. Poole and C. F. Poole, J. Planar Chromatogr. 5 (1992) 200.
[206] B. Spangenberg and K.-F. Klein, J. Chromatogr. A 898 (2000) 265.
[207] B. Spangenberg, B. Ahrens and K.-F. Klein, Chromatographia 53 (2001) S-438.
[208] I. Vovk and M. Prosek, J. Chromatogr. A 779 (1997) 329.
[209] M. Prosek, A. Golc-Wondra and I. Vovk, J. Planar Chromatogr. 14 (2001) 100.
[210] J. A. Cosgrove and R. B. Bilhorn, J. Planar Chromatogr. 2 (1989) 362.
[211] I. Vovk amd M. Prosek, J. Chromatogr. A 768 (1997) 329.
[212] S. Ebel and T. Henkel, J. Planar Chromatogr. 13 (2000) 248.
[213] S. Essie and K.-A. Kovar, Chromatographia 53 (2001) 321
[214] E. Hahn-Deinstrop, A. Koch and M. Muller, J. Planar Chromatogr. 11 (1998) 404.
[215] J. Summanen, T. Yrjonen, R. Hiltunen and H. Vuorela, J. Planar Chromatogr. 11 (1998) 421.
[216] S. Mustoe and S. McCrossen, J. Planar Chromatogr. 14 (2001) 252.
[217] S. P. Mustoe and S. McCrossen, Chromatographia 53 (2001) S-474.
[218] G. W. Somsen, W. Morden and I. D. Wilson, J. Chromatogr. A 703 (1995) 613.
[219] G. W. Somsen, P. G. J. H. ter Riet, C. Gooijer, N. H. Velthorst and U. A. Th Brinkman, J. Planar Chromatogr. 10 (1997) 10.
[220] A. G. Bhullar, D. A. Karlsen, K. Backer-Owe, K. Le Tran, E. Skalnes, H. H. Berchelmann and J. E. Kittelsen, J. Petrol. Geol. 23 (2000) 221.

[221] C. Weins and H. Jork, J. Chromatogr. A 750 (1996) 403.
[222] I. K. Rhee, M. van de Meent, K. Ingkaninan and R. Verpoorte, J. Chromatogr. A 915 (2001) 217.
[223] G. Ebertz, H.-G. Rast, K. Burger, W. Kreiss and C. Weisemann, Chromatographia 43 (1996) 5.
[224] T. Reemtsma, Anal. Chim. Acta 426 (2001) 279.
[225] S. Nagy, B. Kocsis, T. Koszegi and L. Botz, J. Planar Chromatogr. 15 (2002) 144.
[226] L. Williams and O. Bergersen, J. Planar Chromatogr. 14 (2001) 318.
[227] I. D. Wilson, E. R. Adlard, M. Cookee and C. F. Poole, *Encyclopedia of Separation Science*, Academic Press, London, 2000.
[228] H. Filthuth, J. Planar Chromatogr. 2 (1989) 198.
[229] J. Szunyog, E. Mincsovics. I. Hazal, and I. Klevovich, J. Planar Chromatogr. 11 (1998) 25.
[230] M. Yoshioka, H. Araki, M. Kobayashi, F. Kaneuchi, M. Seki, T. Miyazaki, T. Utsuki, T. Yaginuma and M. Kakano, J. Chromatogr. 507 (1990) 221.
[231] O. Klein and T. Clark, J. Planar Chromatogr. 6 (1993) 369.
[232] E. Mincsovics, B. D. Kiss, G. Morovjan, K. B. Nemes and I. Kiebovich, J. Planar Chromatogr. 14 (2001) 312.
[233] P. Vingler, C. Gerst, N. Boyera, I. Galey, C. Christelle, B. A. Bernard, T. Dzido, F. Tardieu, C. Hennion, H. Filthurth and G. Charpak, J. Planar Chromatogr. 12 (1999) 244.

# Chapter 7

# Supercritical Fluid Chromatography

## 7.1 INTRODUCTION

Supercritical fluid chromatography (SFC) has been seeking a permanent home in analytical laboratories since its inception in the early 1960s [1-10]. During this time it has enjoyed peaks of passing popularity championed as an extension of gas chromatography or a replacement for high-pressure liquid chromatography. In reality it is neither of these and a bit of both. Supercritical fluid chromatography is able to extend the molecular weight separating range of gas chromatography at the expense of some loss in efficiency and longer separation times. In the late 1980s supercritical fluid chromatography was reborn largely as an open tubular column technique using column and detector technology imported from gas chromatography and mobile phase delivery and injection systems from liquid chromatography [1-5]. The compromises were generally unsatisfactory. The small optimal column internal diameters combined with their low flow rates impeded instrument development and the reliance on carbon dioxide as virtually the only convenient mobile phase for use with flame-based detectors restricted applications to compounds with adequate solubility in carbon dioxide, a relatively non-polar solvating medium. Open tubular column supercritical fluid chromatography found a niche for itself in the separation of (generally) thermally stable compounds of low polarity (such as hydrocarbon-based and siloxane-based polymers, petroleum products, fats and oils, etc). These compounds were difficult to separate by gas chromatography because extreme temperatures were required to obtain adequate volatility. Simultaneously it acquired a reputation for being unsuitable for the separation of polar compounds because of poor peak shapes and, in some cases low mass recovery. Perhaps the main failing at this time was that excessive exuberance had raised expectations of the general capability of open tubular column supercritical fluid chromatography too high and using open tubular columns of the desired dimensions with available instrumentation was too problematic for entry into quality control laboratories. Contemporary use of open tubular column supercritical fluid chromatography has continued to decline during the 1990s with little expectations of an imminent resurgence of interest outside its niche application areas.

Supercritical fluid chromatography was reincarnated in the 1990s as a packed column technique, primarily as a replacement for normal-phase liquid chromatography, and mostly for the separation of polar compounds [6-10]. Uptake has been slow but steady during the last decade, faced with reservations based on poor experiences with open tubular column supercritical fluid chromatography and confusion over the new and contrary direction, seemingly against the accepted view of the limitations of supercritical fluid chromatography. This change in ideology was fostered by the movement to instrumentation more like that used in liquid chromatography, improved column packing materials, the general use of binary mobile phases containing additives to control undesirable column interactions, and the switch to spectrophotometric detectors with high-pressure cells for routine applications. Most modern instruments use reciprocating pumps specifically designed for compressible fluids and provide for independent pressure, flow and temperature control. They also allow accurate

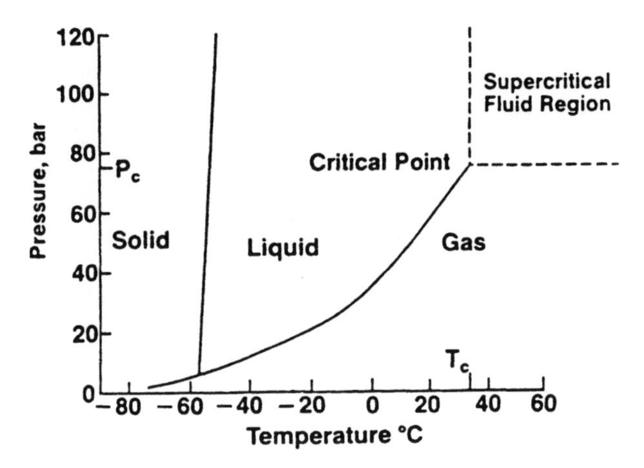


Figure 7.1. Phase diagram for carbon dioxide

mixing of compressible fluids and liquids, which is essential now that composition variations rather than density variations are the most widely used approach for method development. Fluid-solvent mixtures are far less compressible than neat supercritical fluids ameliorating some of the problems of efficiency loss and mobile phase velocity and retention heterogeneity for columns containing small particles and/or of longer lengths.

A supercritical fluid is simply an element or compound at a temperature above its critical temperature that is simultaneously compressed to a pressure exceeding its critical pressure, Figure 7.1. The supercritical fluid region is not a separate state of matter but simply an extension of the gas and liquid phase regions to a temperature and pressure domain where phase separation is no longer possible. The transition of physical properties from the gas or liquid phase to the supercritical fluid region is smooth and not marked by the dramatic changes that generally accompany transitions at phase boundaries, such as the melting of a solid or boiling of a liquid. The properties of supercritical fluids (e.g. density, viscosity, diffusivity, etc.) are continuously variable by changing the temperature and/or pressure above the critical point. The fluid may exhibit gas-like and liquid-like properties at the extremes of its range represented by the dotted lines in Figure 7.1 and some combination of these properties, at other locations that depend on the temperature and pressure coordinates. Table 7.1 provides some representative data for the density of carbon dioxide at different temperatures above the critical temperature at 72 atmospheres (close to the critical pressure) and 400 atmospheres (close to the highest pressure normally used in supercritical fluid chromatography). For carbon dioxide virtually the whole density range from about 0.1-1.0 g / ml is easily accessible representing a wide range of fluid properties.

Typical properties of a gas (helium) under gas chromatographic conditions, a liquid (water) under liquid chromatographic conditions, and carbon dioxide under low and high density conditions, as might be used in supercritical fluid chromatography, are

Table 7.1
Change in density of supercritical fluid carbon dioxide with pressure and temperature

| Temperature (°C) | Pressure (atm) | Density (g/ml) |
|---|---|---|
| 40 | 72 | 0.22 |
| | 400 | 0.96 |
| 60 | 72 | 0.17 |
| | 400 | 0.90 |
| 80 | 72 | 0.14 |
| | 400 | 0.82 |
| 100 | 72 | 0.13 |
| | 400 | 0.76 |
| 120 | 72 | 0.12 |
| | 400 | 0.70 |
| 140 | 72 | 0.11 |
| | 400 | 0.64 |

Table 7.2
Representative properties of typical chromatographic mobile phases

| Mobile phase | Temperature (°C) | Pressure (atm) | Density (g/ml) | Diffusivity ($cm^2/s$) | Viscosity (cP) |
|---|---|---|---|---|---|
| Helium | 200 | 1.5 | $2x10^{-4}$ | 0.1-1 | 0.02 |
| Carbon dioxide | | | | | |
| low density | 100 | 80 | 0.15 | $10^{-3}$ | 0.02 |
| high density | 35 | 200 | 0.8 | $10^{-4}$ | 0.1 |
| Water | 20 | | 1.0 | $10^{-5}$ | 1.0 |

summarized in Table 7.2 [11]. The capacity of a mobile phase for intermolecular interactions is related to its density. Supercritical fluids have densities generally between those of gases and liquids but closer to those of liquids. They are expected to have higher solubilizing power than gases but are not as effective solvents as liquids. Unlike liquids, the density and solvating power of a supercritical fluid can be varied at a constant temperature by changing its pressure. Under normal conditions liquids are virtually incompressible and the external pressure is irrelevant to their solvent properties.

Diffusion coefficients in supercritical fluids are intermediate between those of gases and liquids but generally closer to those of liquids. They may be an order of magnitude greater than in liquids, however, which has important implications for chromatography affecting separation times, column characteristics and instrument design. Typical minimum plate heights for packed column supercritical fluid chromatography and liquid chromatography are similar. The most important difference is that the minimum plate height for supercritical fluid chromatography is achieved at linear velocities 5 to 10 times higher than for liquid chromatography, and separation times are reduced. Also, the resistance to mass transfer term in supercritical fluid chromatography is not as large as it is for liquids, which allows a further increase in the separation speed without a significant decrease in efficiency. It should always be possible to obtain faster separations by supercritical fluid chromatography than by liquid chromatography but these separation speeds do not approach those of gas chromatography, even for low-density supercritical fluids.

The rate of radial diffusion controls the dimensions of open tubular columns when reasonable efficiency per unit column length is required. Open tubular columns need only have internal diameters of about 0.1-0.5 mm in gas chromatography to obtain reasonable efficiency, conditions that are easily met. For equivalent efficiency in liquid chromatography column internal diameters should be less than about 0.01 mm. These columns are neither easy to prepare nor simple to operate, and are rarely used in practice. The more favorable diffusion properties of supercritical fluids allow larger diameter open tubular columns to be used, about 0.05-0.10 mm, with relaxed instrument constraints compared to liquid chromatography. Thus, both packed and open tubular columns are used in supercritical fluid chromatography, where each type has distinct advantages, while for liquid chromatography packed columns are the only practical alternative. Since the solute diffusion coefficients vary strongly with fluid density, open tubular columns of typical dimensions will provide more favorable separation properties at low mobile phase densities. At high densities the column dimensions required for optimum performance approach those of liquid chromatography.

The viscosities of supercritical fluids are intermediate between those of gases and liquids, but are closer to those of gases. For a fixed column pressure drop longer columns or higher flow rates are possible in supercritical fluid chromatography compared with liquid chromatography. Supercritical fluids, however, are highly compressible and a large column pressure drop has the effect of decreasing the density along the length of the column and increasing the local mobile phase velocity leading to additional band broadening. For the same column pressure drop, the greater diffusivity of gases means that supercritical fluid chromatography cannot approach gas chromatography either in the total plate number that can be achieved or separation speed.

## 7.2 MOBILE PHASES

The properties desired of a supercritical fluid for use in chromatography are low critical constants, low chemical reactivity, low toxicity and flammability, availability in a high purity grade at a reasonable cost, and compatibility with common gas and condensed phase detectors. Only carbon dioxide meets these requirements for the most part. The critical properties of some other supercritical fluids are summarized in Table 7.3 [3,12]. Nitrous oxide has similar solvent strength to carbon dioxide but its strong oxidizing properties are a hazard [13]. Sulfur hexafluoride is a weaker eluent than carbon dioxide and difficult to obtain in adequate purity. Xenon has low solvent strength, is expensive, but has found some applications when combined with FTIR detection due to its favorable spectral transparency extending from the vacuum UV to the NMR region [14]. Low molecular mass n-alkanes are weak solvents, have relatively high critical constants, and cannot be used with flame-based detectors [3,15]. These fluids are flammable and require extra care when used.

Table 7.3
Physical properties of some supercritical fluids

| Fluid | Critical Parameters | | | Density at 400 atm (g/ml) |
|---|---|---|---|---|
| | Temperature (°C) | Pressure (atm) | Density (g/ml) | |
| Carbon dioxide | 31.3 | 72.9 | 0.47 | 0.96 |
| Nitrous oxide | 36.5 | 72.5 | 0.45 | 0.94 |
| Sulfur hexafluoride | 45.5 | 37.1 | 0.74 | 1.61 |
| Xenon | 16.6 | 58.4 | 1.10 | 2.30 |
| Butane | 152.0 | 37.5 | 0.23 | 0.50 |
| Pentane | 196.6 | 33.3 | 0.23 | 0.51 |
| Dichlorodifluoromethane | 111.8 | 40.7 | 0.56 | 1.12 |
| Trifluoromethane | 25.9 | 46.9 | 0.52 | 1.15 |
| 1,1,1,2-Tetrafluoroethane | 101.2 | 40.2 | * | |
| Ammonia | 132.5 | 112.5 | 0.24 | 0.40 |
| Water | 374.4 | 226.8 | 0.34 | |
| Methanol | 240.5 | 78.9 | 0.27 | |
| Acetonitrile | 274.7 | 47.7 | 0.24 | |

* Density = 1.147 at 75°C and 200 bar and 0.985 at 125°C and 200 bar.

All the above fluids are weak mobile phases [16,17]. Carbon dioxide has similar solvent properties to liquid hydrocarbons with the principal intermolecular interactions being dispersion and induction, since carbon dioxide has no dipole moment of its own. Polar substances capable of a wider range of intermolecular interactions tend to have (excessively) high critical constants incompatible with the stable operation of contemporary columns and instruments (see Table 7.3). Of the potentially useful polar fluids the chlorofluorocarbons and hydrofluorocarbons have favorable critical constants and modest dipole moments, but availability, cost and environmental concerns have limited their use [18,19]. The dipolar and hydrogen-bond basic hydrofluorocarbon refrigerant 1,1,1,2-tetrafluoroethane (HFC-134a) has a reasonable potential for future use [20]. Ammonia is a strong solvent but is too aggressive and toxic for practical use [21,22]. It dissolves silica-based materials and corrodes instrument components.

Carbon dioxide is available in a number of grades, including a purified supercritical fluid chromatography grade, and is sold in pressurized cylinders in which the bulk of the carbon dioxide is in the liquid state. Cylinders with a helium headspace are available to assist in dispensing liquid carbon dioxide without having to cool pump heads, etc. These cylinders are unsuitable for use in supercritical fluid chromatography when reproducible retention is important [23-25]. Under typical conditions the concentration of helium dissolved in the liquid carbon dioxide varies from about 0 to 5 mole percent. The lower elution strength of carbon dioxide-helium mixtures compared with carbon dioxide alone results from the significant reduction in the density of the mixed mobile phase and variation in retention is related to changes in the composition of the dispensed carbon dioxide-helium mixture as the cylinder contents are consumed. Cylinder lubricants, a mixture of chlorotrifluoroethylene oligomers, have been identified as trace contaminants in both high and low purity grades of carbon dioxide [26]. Primary

| te | 2025/06/11 |
| --- | --- |
| ne | 10:22:49 |
| inter | OCE-2 |
| **ror identification** | ioerror |
| **fending Command** | --image-- |
| **erand Stack** | Top -dict- |